CLEAN
FUEL SUPPLY

This book is to be returned on or before
the last date stamped below.

27 NOV 1990

15 MAR 1991

30 JAN 1991

FOR ECONOMIC CO-OPERATION AND DEVELOPMENT
PARIS 1978

The Organisation for Economic Co-operation and Development (OECD) was set up under a Convention signed in Paris on 14th December 1960, which provides that the OECD shall promote policies designed:
- to achieve the highest sustainable economic growth and employment and a rising standard of living in Member countries, while maintaining financial stability, and thus to contribute to the development of the world economy;
- to contribute to sound economic expansion in Member as well as non-member countries in the process of economic development;
- to contribute to the expansion of world trade on a multilateral, non-discriminatory basis in accordance with international obligations.

The Members of OECD are Australia, Austria, Belgium, Canada, Denmark, Finland, France, the Federal Republic of Germany, Greece, Iceland, Ireland, Italy, Japan, Luxembourg, the Netherlands, New Zealand, Norway, Portugal, Spain, Sweden, Switzerland, Turkey, the United Kingdom and the United States.

* * *

© OECD, 1978
Queries concerning permissions or translation rights should be addressed to:
Director of Information, OECD
2, rue André-Pascal, 75775 PARIS CEDEX 16, France.

TABLE OF CONTENTS

PREFACE	5
CONCLUSIONS	7
General	7
Desulphurisation Techniques	9
OECD Europe	10
OECD North America	12
Japan	12

Chapter I

INTRODUCTION	15

Chapter II

SO_2 EMISSIONS LEVELS IN THE OECD - 1974 AND 1985	17
Tables	18
Japan	30
OECD Europe	35
OECD North America	40

Chapter III

POTENTIAL FOR REDUCTION OF SO_2 EMISSIONS	46
Supply of Naturally Low Sulphur Fuels	46
Coal Cleaning - North America	48
Coal Cleaning - OECD Europe	55
Residual Fuel Oil Desulphurisation - Japan	58
Residual Fuel Oil Desulphurisation - OECD Europe	60
Residual Fuel Oil Desulphurisation - North America	61
Flue Gas Desulphurisation - OECD Europe - Power Plants	62
Flue Gas Desulphurisation - OECD North America - Power Plants	67
Flue Gas Desulphurisation - Japan - Power Plants	73
Flue Gas Desulphurisation - OECD - Industrial Combustion	76

Chapter IV

ECONOMICS OF DESULPHURISATION	77
Residual Oil Combustion/Residual Oil Desulphurisation	78
Residual Oil Combustion/Flue Gas Desulphurisation	87
Residual Oil Combustion/Segregation of High-Sulphur Fuel Oil to FGD	92
Coal Combustion/Coal Cleaning	93
Coal Combustion/Flue Gas Desulphurisation	96
Coal Combustion/Coal Cleaning vs. Flue Gas Desulphurisation	98
Summary	100
BIBLIOGRAPHY	102

PREFACE

The availability of low sulphur fuels and the introduction of desulphurisation technologies will play a major role in the formulation of international control strategies on the general sulphur problem.

This study, which addresses energy and environmental policy makers and air pollution managers, analyses the technological and economic factors in the OECD regions up to the mid 1980s. The report provides forecasts of emission levels, costs of desulphurisation, and the potentials for reduction. The analysis shows a potential for further reducing sulphur emissions in North America and Europe without severe technical or economic bottlenecks.

As one element in the efforts to reduce the environmental impact for energy production and use, the OECD Council has adopted recommendations related to these issues in 1974 and 1976. This report and the conclusions drawn from it were adopted by the Environment Committee in April 1978 and form the basis for future policies and agreements on sulphur control.

CONCLUSIONS

On the 22nd October 1976, concerning the reduction of environmental impacts from energy production and use, the Council, on the proposal from the Environment Committee, recommended /C(76)162(Final)/ that Member Countries participate in an international examination of control strategies needed to reduce emissions of sulphur compounds to acceptable levels in view of their effects on man and the environment. This report provides the technological and economic assessment of control strategies in the light of future energy supply and sulphur oxides emission forecasts.

The most probable forecast of 1985 SO_2 emissions for OECD as a whole is 57 million metric tons, a 23 per cent increase over the 1974 levels.

The policies for reducing this increase in SO_2 emissions are:

a) increasing energy conservation and developing appropriate indigenous energy resources,
b) increasing the share of imports of low sulphur crude oil, within practical limits, and
c) increasing the installed capacity of sulphur removal technologies.

On the 14th November 1974, concerning measures required for further air pollution control, the Council, on the proposal of the Environment Committee at its First Meeting at Ministerial Level, recommended /C(74)219/ that the Governments of Member countries should strive with all practicable speed to reduce emissions of sulphur oxides and particulate matter

i) by applying the best available abatement techniques for these pollutants;
ii) by encouraging the expanded and more efficient production as well as the more effective use of clean fuels.

There are no technological barriers hindering the implementation of the Council recommendation and the economic implications are shown below.

The report concludes that:

a) Governments can take actions today to promote the application of the best available technologies for sulphur reduction, but there are supply limitations to increased imports of low sulphur oils. Additionally, political obstacles in the producing countries may hinder the promotion of higher imports of low sulphur oil.

b) In the present circumstances, the additional cost of purchasing low sulphur oil (the sulphur premium), the additional cost of desulphurising high sulphur residual oil and the additional cost of installing FGD on large, new oil-fired power plants are all of the same order of magnitude.

c) If governments were to adopt policies which maintain or increase the sulphur premium for oil over the long term (beyond 1985), the economic incentive for using desulphurisation technologies as an alternative to importing low sulphur fuels would be established.

d) Maximum use of coal washing to reduce sulphur is advocated. Where required, coal washing should be combined with flue gas desulphurisation.

e) Member countries should carry out a more detailed analysis of their own situation and future options for sulphur emissions reductions within the framework established by this report.

f) In North America and Japan, the governments have developed policies which encourage or stipulate the use of desulphurisation technologies. In Europe, many countries are now considering what national policies and international agreements will be needed. In OECD Europe there is a great potential for reducing sulphur emissions without severe technical or economic bottlenecks.

g) Through the use of desulphurisation technologies, the emissions of 25×10^6 metric tons of SO_2 per year in the 1985 "worst" case could be reduced to approximately $9-10 \times 10^6$ metric tons at an annual operating cost of $5 billion per annum. This would represent about 0.3 per cent of OECD European GDP or, on a per capita basis, $15 per annum. If a minor reduction to the 1974 level of 19×10^6 metric tons per year were acceptable, the cost would be reduced to $1.8 billion per annum representing about 0.1 per cent of OECD European GDP or, on a per capita basis, $5 per annum.

h) Comparison of the costs of desulphurisation strategies is necessarily on the basis of total annual costs. Implementation of these strategies involves large capital investment

programmes. Such investments, involving billions of
dollars, will have significant impacts on many sectors of
the economy and need to be integrated into countries'
economic planning policies. The financing of these investments,
their impact on the iron and steel, heavy engineering
and construction sectors, employment opportunities,
trade and balance of payments are all important factors
to be considered by Member countries.

Emission Forecasts

The study shows that, for OECD as a whole, total SO_2 emissions from fuel combustion in 1974 were about the same as in 1968, despite a 26 per cent increase in fuel consumption.

The SO_2 emissions from fuel combustion were forecast for 1985 using the energy forecasts from the "World Energy Outlook"(2) and assuming no increase of desulphurisation capacity over that already installed and planned. In the "worst" case there could be an increase of about 23 per cent over the 1974 levels. In the "best" case, assuming that countries strive towards their energy independence objectives, there could be a decrease of the order of 6 per cent or some 2.7 million metric tons of SO_2. These changes in energy policy can be seen to be critical.

The exact position for the different OECD regions is shown in the table below.

SUMMARY OF ESTIMATED AND FORECAST SO_2 EMISSIONS IN THE OECD, 1968 TO 1985

10^6 metric tons of SO_2

	1968	1974	1985 Reference Case	1985 Accelerated Policy Case
OECD Europe	16.7	19.7	22.1-25.4	19.6-21.4
North America	26.6	24.2	24.9-28.1	22.4-25.7
Japan	4.0	2.4	2.6-3.6	1.7-3.3
Total	47.3	46.3	49.6-57.1	43.7-50.4

If the OECD countries were to achieve the goals of energy conservation and development of indigenous energy resources set out in the Accelerated Policy Case, the sulphur emission standstill trend which has existed in OECD from 1968 to 1974 would probably continue up to 1985. The accelerated policy case will require vigorous measures by Member countries. Failing greatly increased efforts, the reference case would be the more likely outcome. (These forecasts are now being revised.) There are also fears that even certain expectations in the Reference Case may not be realised. Perhaps the most important factor will be the share of total energy that nuclear power is forecast to provide. If nuclear output falls short of projections, there will be a tendency to shift to fossil fuels and a consequent increase in SO_2 emissions. In this situation oil will be the balancing fuel and will of necessity be made up of the medium to high sulphur crude oils.

The SO_2 emissions forecast for 1985 in this study, even in the "worst" case, are much lower than earlier predictions. For example, the OECD "Report and Conclusions of the Joint Ad Hoc Group on Air Pollution from Fuel Combustion in Stationary Sources"(7) forecast SO_2 emissions in all OECD areas to reach a total of 94.6 million metric tons in 1980, a doubling from the 1978 level. In comparison, the current study in the "worst" case forecasts SO_2 emissions in 1985 to be 57.1 million metric tons. The reasons for this difference in forecasts of SO_2 emissions are not hard to find. First, there have been changes in the level and pattern of energy consumption and supply since 1973 when oil prices were quadrupled and supplies restricted.

Secondly, the situation has also been helped by a move towards lower sulphur fuel supplies and the gradual introduction of desulphurisation and stack gas scrubbing techniques, especially true in the case of Japan.

Desulphurisation Techniques

The technology exists for reducing the sulphur content of coal. The extent of removal is dependent on the physical and chemical characteristics of the sulphur in the coal. In regions where coal washing is extensively practised but not for the purpose of maximum sulphur reduction, the potential exists to upgrade this cleaning practice to remove sulphur. The expected incremental cost is ∅1.40-2.20 per metric ton of coal washed (1980 dollars).

In regions where washing of combustion coals is not extensively practised, considerable potential for sulphur removal also exists, especially if coal washing can be combined with flue gas desulphurisation (FGD) in coal-fired power plants. This approach has cost advantages over using flue gas desulphurisation by itself. It involves the segregation of the coal in the washing plant into a clean coal product and a (high sulphur) middlings product, of which the latter would go to the power plants with FGD.

Gas oil desulphurisation is being undertaken in all OECD regions and it is planned that by 1985 gas oil will generally have a sulphur content of around 0.3 per cent. This level represents about the current feasible limit for desulphurisation and therefore little potential exists for further reductions of SO_2 emissions from gas-oil.

The lowest cost strategy for residual oil desulphurisation is to desulphurise the high and medium sulphur residues to 0.5 per cent sulphur. If these desulphurised products are blended with residue naturally low in sulphur and undesulphurised residue into a 1 per cent sulphur fuel oil pool and the cost of the desulphurisation is distributed over this pool, the incremental cost of the 1 per cent sulphur fuel oil will be lower than today's sulphur premium for imported low sulphur crude. In fact, for Europe, up to 7.6×10^6 metric tons of SO_2 can be removed in the 1985 "worst" case at a cost less than the existing 1977 sulphur premium for low sulphur fuel oil.

The costs for residual oil desulphurisation and for flue gas desulphurisation on oil-fired power plants are comparable. The lowest cost strategy for sulphur removal from oil is to segregate the high and medium sulphur residues into two pools, with the highest sulphur pool being used in oil-fired power plants with FGD and the other pool being desulphurised. This minimises the number of power stations which need to install FGD and optimises the cost for both technologies.

It should be noted that the costs of control technologies estimated in the report have associated error bands, generally ranging from 10 per cent to 30 per cent. This must be recognised when viewing the absolute cost to obtain a certain level of emission reduction. However, the main purpose of the costs analysis has been to compare the costs for different technologies and for the same technology applied in different situations. For determining the relationships among these costs, the cost undertainty is of less importance.

<u>OECD Europe</u>

Europe's SO_2 emissions in 1985 are forecast to come one-third from coal combustion and two-thirds from oil combustion, the coal being used for electric power generation and the residual oil being split evenly between the electric power generation and industry sectors. Since there is very little sulphur control technology installed or planned in Europe, the 1985 emissions will be extremely sensitive to the energy policies of European countries and fuel exporting countries.

The study shows the sensitivity of the future emissions to the oil import and export policies in Europe. The lowest emissions will result if Europe consumes all of its own North Sea oil production, which is low in sulphur, and continues in 1985 to import the same quantity of low sulphur crude oil as in 1974. The likelihood of this will be dependent upon crude oil prices, refinery structure, product demands, national energy policies and other factors. However, it does seem unrealistic to assume that all the North Sea oil will be used in Europe without reduction of current imports of low sulphur oil.

It is therefore reasonable to conclude that the 1985 emissions will approach the high end of the range forecast for the 1985 Reference Case.

From the analysis of control technologies and their costs, the following European approach for maximum reduction of total SO_2 emissions would have the lowest cost of sulphur removal (1980 dollars):

- segregate high sulphur and medium sulphur residual fuel oils,
- install flue gas desulphurisation in all oil-fired power plants over 200 MW and constructed since 1974 and use the high sulphur fuel oils ($475-630/metric ton of S removed),
- desulphurise by direct residue desulphurisation the remaining high and medium sulphur residual fuel oils to a level of 0.5 per cent S ($630-810/metric ton of S removed),
- physically wash all hard coals to minimise sulphur content ($560-2100/metric ton S removed),
- install FGD on all lignite-fired boilers over 100 MW and constructed since 1967 ($520-890/metric ton S removed),
- require that all imported coals be washed to minimise sulphur content, and
- use naturally low sulphur or cleaned fuel in the domestic, commercial and small industrial sector where FGD is not practical.

Using this approach, it would be possible to reduce emissions of SO_2 in the 1985 "worst" case by about 16 million metric tons to a level slightly under 10 million metric tons. However, from a practical point of view, it would be difficult to implement such an approach by 1985.

To maintain emissions at their present level of 20 million metric tons of SO_2, a standstill approach could be put into practice by 1985. In the "worst" case this would mean a reduction of 6×10^6 metric tons SO_2. The washing of all hard coals could reduce SO_2 emissions by 1×10^6 metric tons at an annual operating cost of approximately $0.35 billion. Installation of FGD on all new (post 1974) lignite-fired boilers over 100 MW would reduce SO_2 emissions by another 1×10^6 metric tons at $0.3 billion in 1985. The remaining 4×10^6 metric tons reduction could be accomplished by segregation of 4.0 per cent S fuel oil to new power plants (post 1980) with FGD or by direct desulphurisation of high sulphur residual oil to 0.5 per cent S. (The cost of low sulphur fuel oil would be incremented by $7/metric ton). The 1985 operating cost would range from $1.0-1.25 billion for these two options. The SO_2 emissions could be reduced by only 3×10^6 metric tons by the purchase of additional low sulphur imported crude oil at an additional cost of about $1.75 billion.

A standstill approach which would require the removal of about 6×10^6 metric tons of SO_2 in 1985 would result in a 1985 annual operating cost of $1.65-1.90 billion.

OECD North America

The study shows that the potential SO_2 emissions in North America in 1985 will come primarily from coal combustion in power stations. However, since the installed and planned flue gas desulphurisation capacity in North America is large (50-80 GW), much of these potential emissions from power stations will be controlled. North America must achieve its planned FGD capacity to avoid an increase of emissions of 20-40 per cent above the forecast.

As a result of this control technology and the fact that coal is domestically produced, the 1985 emissions in North America will be much less sensitive to the energy and fuel import policies than will be the emissions in Europe. The range shown in the summary table reflects only the uncertainty in the amount of flue gas desulphurisation to be installed.

North America is the one OECD region where coal cleaning represents a method for substantial SO_2 emission reduction by 1985. This is due in part to the expected increase in coal use by 1985 and also to the fact that currently only a relatively small proportion of production for combustion is washed.

The segregation of coal in the washing plant into high and low sulphur fractions is feasible. This would permit higher sulphur coals and middlings from the washing plant to be used in the large number of power plants with flue gas desulphurisation. The low sulphur coal could be used to encourage industry to use more coal and less oil and gas. It is estimated that 500 million metric tons of coal averaging 1.0 per cent sulphur content could be segregated out of the total 857 million metric tons of combustion coal.

Japan

In Japan the problem of SO_2 emissions is almost exclusively associated with the use of residual fuel oil, with the industrial sector as the main emitter. Japan has extensively installed technologies for residual fuel oil desulphurisation and flue gas desulphurisation. However, as the study shows, there is some sensitivity of the future emission levels to the energy and fuel policies of Japan and of other countries which compete for or export the low sulphur fuel. If the 1985 emissions do approach the Reference Case with the highest emissions, there will be scope for installing more residual oil desulphurisation capacity to remove an additional 1.0 million metric tons of SO_2. In all other cases, to remove additional sulphur more sophisticated or more efficient technologies or operating techniques would be required.

Chapter I

INTRODUCTION

The report addresses the problem of limited supplies of clean fuel. With increasing energy consumption, the supply of clean fuel may not be sufficient to meet future levels of demand as more countries seek to reduce emissions of sulphur oxides. The term clean fuel is here used to refer to fuels from which there are low emissions of sulphur oxides, principally sulphur dioxide (SO_2). These can either be naturally low in sulphur, or be desulphurised prior to combustion. If the factors which affect this supply can be identified, it may be possible for governments to take action where required, individually and collectively, to increase the supply of clean fuels and fuel cleaning techniques. It will also be possible to determine in which cases to desulphurise the gaseous combustion products thus allowing high sulphur fuels to have low final emissions of SO_2.

The question of whether any action on clean fuel supplies is required, what form it should take and when is a matter for discussion. It is hoped that this report, in providing information on current SO_2 emissions, forecasts of future levels and an analysis of how and at what cost these levels can be further reduced, will play an important role in such discussions.

All OECD countries have recognised the need to attain acceptable ground level concentrations of SO_2 and many have implemented ambient air quality standards and/or emission standards for SO_2. The transport of SO_2 across frontiers has also concerned many countries. These factors have led all OECD Member countries to agree to Council Recommendation C(74)219, to strive with all practicable speed to reduce emissions of sulphur oxides.

In addition, the recent OECD study on long range transport of air pollutants (1) has confirmed that sulphur compounds do travel long distances (several hundred kilometres or more) in the atmosphere and has shown that the air quality in any one country is measurably affected by emissions from its neighbouring countries and many of the other countries in the continent. The overall solution to the problem of SO_2 will only be found through international co-operation in which national policies are implemented to attain acceptable ambient air quality at the same time as to minimise transport across frontiers.

Because of the importance of trade in energy supplies and the problem of long range transport of sulphur compounds, it is appropriate that the question of clean fuel be examined from an international viewpoint. This report analyses the situation in the three geographical regions of OECD, namely, Europe, North America and Japan. (Australia and New Zealand have not been included at present. However, preliminary calculations show that total SO_2 emissions from fuel combustion in these countries are relatively low, although these may be concentrated in certain areas.) The technologies for reducing SO_2 emissions in many cases cause secondary pollution in other media. Technologies exist for control of these secondary pollutants **and** are included in the costs.

The time frame for this report is 1985. This recognises the fact that any decisions taken on clean fuel supply or fuel cleaning techniques would not have any major impact until the mid 1980s because of the long investment lead times involved. Only these technologies that have already been commercially developed are considered as being available for wide-scale application for 1985. Therefore the report does not include such technologies as liquefaction, chemical cleaning of coal or fluidised bed combustion. These technologies may, however, be significant when dealing with energy strategies and sulphur oxide control to the year 2000.

The energy forecasts by fuel type for 1985 were taken from a recent OECD study, "World Energy Outlook".(2) This report therefore provides a sulphur oxide environmental assessment of the World Energy Outlook.

Chapter II

SO_2 EMISSION LEVELS IN THE OECD - 1974 AND 1985

To provide a basis for investigating the factors which affect the supply of low sulphur fuels and the introduction of sulphur removal technologies, sulphur dioxide emissions from fuel combustion in stationary sources have been estimated for 1974 using published OECD energy statistics,(3) (4) (5) and (6) and knowledge of fuel sulphur content and user sector emissions factors. (Tables 1, 4, 7 and 10.)

The recent World Energy Outlook (2) provides the basis for the forecasts of sulphur dioxide emissions for 1985. (Tables 2, 3, 5, 6, 8, 9, 11 and 12.) Two different scenarios are taken, the **Reference Case** which assumes a continuation of present policies governing supply expansion and conservation by OECD Member countries, and the **Accelerated Policy Case** which reflects additional policy options which could be taken in the fields of conservation, increased production of indigenous energy sources, the increased use of natural gas, and the increased penetration of nuclear power (at the expense of new coal and oil-fired power stations). The World Energy Outlook, however, was not the source of the complete sectoral dissection of energy consumption shown in the tables.

The fuels considered are hard coals, patent fuels and non-metallurgical coke, brown coals, brown coal briquettes, gas/diesel oils used for burning (but not motor use), and residual fuel oil. The major end-uses covered include not only stationary source combustion of fossil fuels but also the transformation of solid fuels to secondary energy sources other than electricity - coke and briquette production. The transport sector is not included.

The sources used for the sulphur contents of fuels were the OECD Report and Conclusions of the Joint Ad Hoc Group on Air Pollution from Fuel Combustion in Stationary Sources,(7) which gave emission factors for 1968, reports from countries, (8) (10) (11) (14) (15) (16) and (19) and information provided by private communication.

In the case of coals and lignites the retention of sulphur in the ash during combustion is also important. For hard coal it is assumed that there is a retention of 5 per cent of the sulphur in the

Table 1

FUEL COMBUSTION IN STATIONARY SOURCES AND SO_2 EMISSIONS IN JAPAN, 1974

(References (3), (4) and (5))

	Hard Coal	Patent Fuel & Coke	Lignite	Lignite Briq.	Crude Oil	Gas/Diesel Oil	Residual Fuel Oil
Transformation:							
Power plants	7.340						44.927
Coke/briquettes	67.600						
Energy sector	0.850	0.200					6.105
Industry	2.010	1.400			20.000	9.798	32.707
Domestic/Commercial	1.200	5.888	0.100			6.791	11.748
Total (excl. Transport)	79.000	7.488	0.100		20.000	17.207	95.487
10^6 Metric Tons Fuel							
Transformation:							
Power plants	0.100						1.797
Coke/briquettes							
Energy sector	0.012	0.002			0.240	0.004	0.244
Industry	0.027	0.016				0.059	1.308
Domestic/Commercial	0.016	0.071	0.001			0.041	0.470
Potential emissions	0.155	0.089	0.001		0.240	0.104	3.819
Fuel desulphurisation							-1.580(1)
10^6 Metric Tons SO_2							

FGD: -0.39   Total Emissions: 2.44

1) Installed HDS capacity.

Table 2

FORECAST FUEL COMBUSTION IN STATIONARY SOURCES AND SO_2 EMISSIONS IN JAPAN, 1985 REFERENCE CASE

		Hard Coal	Patent Fuel & Coke	Lignite	Lignite Briq.	Crude Oil	Gas/Diesel Oil	Residual Fuel Oil
10^6 Metric Tons Fuel	Transformation:							
	Power plants	19.7				20.0	0.8	35.3
	Coke/briquettes	90.0						
	Energy sector	0.7	0.3					11.8
	Industry	2.0	1.4				22.2	101.2
	Domestic/Commercial		5.9				15.3	11.7
	Total (excl. Transport)	112.4	7.6				38.3	160.0
10^6 Metric Tons SO_2	Transformation:							A B (1)
	Power plants	0.27				0.24	0.01	1.98 1.69
	Coke/briquettes							
	Energy sector	0.01	0.02					0.66 0.57
	Industry	0.03	0.07				0.13	5.67 4.86
	Domestic/Commercial						0.09	0.66 0.56
	Potential emissions	0.31	0.09			0.24	0.23	8.97 7.68
	Fuel desulphurisation							−4.40 −4.20 (2)

FGD: A: −1.80 B: −1.80   Total Emissions: A: 3.64 B: 2.55

1) Dependent on crude supply (see text).
2) 1985 HDS capacity.

Table 3

FORECAST FUEL COMBUSTION IN STATIONARY SOURCES AND SO_2 EMISSIONS IN JAPAN, 1985 ACCELERATED POLICY CASE

	Hard Coal	Patent Fuel & Coke	Lignite	Lignite Briq.	Crude Oil	Gas/Diesel Oil	Residual Fuel Oil
Transformation:							
Power plants	19.7				20.0	0.6	20.3
Coke/briquettes	90.0						
Energy sector	0.7	0.3					10.2
Industry	2.0	1.4				19.0	94.8
Domestic/Commercial		5.9				13.3	11.7
Total (excl. Transport)	112.4	7.6				32.9	137.0
Transformation:							A B (1)
Power plants	0.27				0.24		1.14 0.61
Coke/briquettes							
Energy sector	0.01	0.02				0.11	0.57 0.31
Industry	0.03	0.07				0.08	5.31 2.84
Domestic/Commercial		0.09				0.19	0.66 0.35
Potential emissions	0.31				0.24		7.68 4.11
Fuel desulphurisation							-3.40 -1.80(2)

10^6 Metric Tons Fuel (top section) / 10^6 Metric Tons SO_2 (bottom section)

FGD: A: -1.80   B: -1.40   Total Emissions: A: 3.31   B: 1.74

1) Dependent on crude supply (see text).
2) 1985 HDS capacity.

Table 4

FUEL COMBUSTION IN STATIONARY SOURCES AND SO_2 EMISSIONS IN OECD EUROPE, 1974

(References (3), (4) and (5))

		Hard Coal	Patent Fuel & Coke	Lignite	Lignite Briq.	Crude Oil	Gas/Diesel Oil	Residual Fuel Oil
10^6 Metric Tons Fuel	Transformation:							
	Power plants	128.907	0.160	128.040	0.952	0.160	1.542	79.962
	Coke/briquettes	119.538		14.056	0.122		0.121	9.137
	Energy sector	3.471	0.538	3.793	0.625		24.539	110.288
	Industry	33.408	5.743	8.178	6.434		108.448	24.761
	Domestic/Commercial	28.775	17.681	3.338	8.133		134.650	224.148
	Total (excl. Transport)	314.099	24.122	157.405		0.160		
10^6 Metric Tons SO_2	Transformation:							
	Power plants	3.204	0.004	1.260	0.008	0.003	0.015	3.998
	Coke/briquettes	0.459		0.004				
	Energy sector	0.084	0.010	0.037	0.001		0.001	0.457
	Industry	0.802	0.104	0.108	0.005		0.245	5.514
	Domestic/Commercial	0.623	0.331	0.069	0.054		1.084	1.238
	Potential emissions	5.172	0.449	1.478	0.068	0.003	1.345	11.207
	Fuel desulphurisation							

FGD: —   Total Emissions: 19.72

Table 5

FORECAST FUEL COMBUSTION IN STATIONARY SOURCES AND SO_2 EMISSIONS IN OECD EUROPE, 1985 REFERENCE CASE

	Hard Coal	Patent Fuel & Coke	Lignite	Lignite Briq.	Crude Oil	Gas/Diesel Oil	Residual Fuel Oil
10^6 Metric Tons Fuel							
Transformation:							
Power plants	165.0		156.0			3.0	157.0
Coke/briquettes	140.0		12.3				
Energy sector	4.0	0.6	4.7	0.1		0.2	12.4
Industry	35.0	5.7	15.7	0.6		36.6	110.3
Domestic/Commercial	16.0	17.7	3.3	6.4		161.7	
Total (excl. Transport)	360.0	24.0	192.0	7.1		201.5	279.7
10^6 Metric Tons SO_2							A B (1)
Transformation:							
Power plants	4.33		2.37				8.48 6.59
Coke/briquettes	0.54						
Energy sector	0.11	0.01	0.07			0.02	0.67 0.52
Industry	0.92	0.10	0.24			0.22	5.96 4.63
Domestic/Commercial	0.42	0.33	0.05	0.05		0.97	
Potential emissions	6.32	0.44	2.73	0.05		1.21	15.11 11.74
Fuel desulphurisation	−0.20(2)						

FGD: A: −0.24 B: −0.24   Total Emissions: A: 25.42 B: 22.05

1) Dependent on crude supply (see text).
2) 1985 cleaning of German coals.

Table 6

FORECAST FUEL COMBUSTION IN STATIONARY SOURCES AND SO$_2$ EMISSIONS IN OECD EUROPE, 1985 ACCELERATED POLICY CASE

	Hard Coal	Patent Fuel & Coke	Lignite	Lignite Briq.	Crude Oil	Gas/Diesel Oil	Residual Fuel Oil
10^6 Metric Tons Fuel							
Transformation:							
Power plants	180.0		156.0			2.0	106.0
Coke/briquettes	140.0		12.3				
Energy sector	4.0	0.6	4.7	0.1		0.1	10.1
Industry	35.0	5.7	15.7	0.6		29.7	110.3
Domestic/Commercial	26.0	17.7	3.3	6.4		131.1	1.2
Total (excl. Transport)	385.0	24.0	192.0	7.1		162.9	227.6
10^6 Metric Tons SO$_2$							A B(1)
Transformation:							
Power plants	4.89		2.37				4.88 3.60
Coke/briquettes	0.54						
Energy sector	0.11	0.01	0.07			0.01	0.47 0.34
Industry	0.95	0.10	0.24			0.18	5.07 3.75
Domestic/Commercial	0.70	0.33	0.05	0.05		0.79	0.06 0.04
Potential emissions	7.19	0.44	2.73	0.05		0.98	10.48 7.73
Fuel desulphurisation	−0.20(2)						

FGD: A: −0.24   B: −0.24   Total Emissions: A: 21.43   B: 19.56

1) Dependent on crude supply (see text).
2) 1985 cleaning of German coals.

Table 7

FUEL COMBUSTION IN STATIONARY SOURCES AND SO_2 EMISSIONS IN THE USA, 1974

(References (3), (4) and (5))

	Hard Coal	Patent Fuel & Coke	Lignite	Lignite Briq.	Crude Oil	Gas/Diesel Oil	Residual Fuel Oil
10^6 Metric Tons Fuel							
Transformation:							
Power plants	342.260		12.973			10.656	71.352
Coke/briquettes	81.820						
Energy sector		2.399					6.540
Industry	58.516	7.222	0.363			8.584	21.581
Domestic/Commercial	10.358					78.130	30.482
Total (excl. Transport)	492.954	9.621	13.336			97.370	129.955
10^6 Metric Tons SO_2							
Transformation:							
Power plants	14.957		0.168			0.064	1.570
Coke/briquettes	0.491						
Energy sector		0.046					0.196
Industry	2.713	0.137	0.005			0.052	0.647
Domestic/Commercial	0.480					0.469	0.914
Potential emissions	18.641	0.183	0.173			0.585	3.327
Fuel desulphurisation							

FGD: −0.156   Total Emissions: 22.753

Table 8

FORECAST FUEL COMBUSTION IN STATIONARY SOURCES AND SO_2 EMISSIONS IN THE USA, 1985 REFERENCE CASE

	Hard Coal	Patent Fuel & Coke	Lignite	Lignite Briq.	Crude Oil	Gas/Diesel Oil	Residual Fuel Oil
10^6 Metric Tons Fuel							
Transformation:							
Power plants	623.1		22.4				72.1
Coke/briquettes	95.0						
Energy sector		2.8					8.9
Industry	122.7	7.2				12.0	65.0
Domestic/Commercial	10.4		0.6			109.5	30.5
Total (excl. Transport)	851.2	10.0	23.0			132.3	176.5
10^6 Metric Tons SO_2							
Transformation:							
Power plants	19.77		0.29				1.59
Coke/briquettes	0.57						
Energy sector		0.05					0.27
Industry	3.89	0.14	0.01			0.07	1.95
Domestic/Commercial	0.33					0.66	0.92
Potential emissions	24.56	0.19	0.30			0.80	4.73
Fuel desulphurisation							

FGD: -4.30 to -7.40   Total Emissions: 23.18 to 26.28

Table 9

FORECAST FUEL COMBUSTION IN STATIONARY SOURCES AND SO$_2$ EMISSIONS IN THE USA, 1985 ACCELERATED POLICY CASE

	Hard Coal	Patent Fuel & Coke	Lignite	Lignite Briq.	Crude Oil	Gas/Diesel Oil	Residual Fuel Oil
10^6 Metric Tons Fuel							
Transformation:							
Power plants	605.6		21.7			7.3	48.8
Coke/briquettes	95.0	2.8					7.7
Energy sector							
Industry	112.7	7.2				10.6	65.2
Domestic/Commercial	10.4		0.6			96.3	30.5
Total (excl. Transport)	823.7	10.0	22.3			114.2	152.2
10^6 Metric Tons SO$_2$							
Transformation:							
Power plants	18.64		0.28			0.04	1.07
Coke/briquettes	0.57	0.05					0.23
Energy sector							
Industry	3.47	0.14				0.06	1.96
Domestic/Commercial	0.32		0.01			0.58	0.92
Potential emissions	23.00	0.19	0.29			0.68	4.18
Fuel desulphurisation							

FGD: -4.30 to -7.40   Total Emissions: 20.94 to 24.04

Table 10

FUEL COMBUSTION IN STATIONARY SOURCES AND SO_2 EMISSIONS IN CANADA, 1974

(References (3), (4) and (5))

		Hard Coal	Patent Fuel & Coke	Lignite	Lignite Briq.	Crude Oil	Gas/Diesel Oil	Residual Fuel Oil
10^6 Metric Tons Fuel	Transformation:							
	Power plants	12.530		3.051			0.093	2.551
	Coke/briquettes	7.526						
	Energy sector	0.011	0.016	0.001			0.020	2.524
	Industry	1.180	0.643	0.299			3.934	6.469
	Domestic/Commercial	0.317	0.003	0.016			12.965	5.285
	Total (excl. Transport)	21.564	0.662	3.367			17.012	16.829
10^6 Metric Tons SO_2	Transformation:							
	Power plants	0.326		0.023			0.001	0.128
	Coke/briquettes	0.050						
	Energy sector	0.001						0.126
	Industry	0.048	0.012	0.002			0.024	0.323
	Domestic/Commercial	0.013					0.078	0.264
	Potential emissions	0.438	0.012	0.025			0.103	0.841
	Fuel desulphurisation							

FGD: —   Total Emissions: 1.419

Table 11

FORECAST FUEL COMBUSTION IN STATIONARY SOURCES AND SO_2 EMISSIONS IN CANADA, 1985 REFERENCE CASE

	Hard Coal	Patent Fuel & Coke	Lignite	Lignite Briq.	Crude Oil	Gas/Diesel Oil	Residual Fuel Oil
10^6 Metric Tons Fuel							
Transformation:							
Power plants	13.8		12.8			0.2	4.3
Coke/briquettes	11.7						3.6
Energy sector							
Industry	4.8	0.6	1.3			5.7	11.1
Domestic/Commercial	0.3					18.6	5.3
Total (excl. Transport)	30.6	0.6	14.1			24.5	24.3
10^6 Metric Tons SO_2							
Transformation:							
Power plants	0.32		0.09				0.22
Coke/briquettes	0.08						0.18
Energy sector							
Industry	0.22	0.01	0.01			0.03	0.56
Domestic/Commercial	0.01					0.11	0.27
Potential emissions	0.63	0.01	0.10			0.14	1.23
Fuel desulphurisation							

FGD: −0.27 to −0.44   Total Emissions: 1.67 to 1.84

Table 12

FORECAST FUEL COMBUSTION IN STATIONARY SOURCES AND SO_2 EMISSIONS IN CANADA, 1985 ACCELERATED POLICY CASE

	Hard Coal	Patent Fuel & Coke	Lignite	Lignite Briq.	Crude Oil	Gas/Diesel Oil	Residual Fuel Oil
10^6 Metric Tons Fuel							
Transformation:							
Power plants	13.8		12.8				0.5
Coke/briquettes	11.7						3.1
Energy sector							
Industry	4.8	0.6	0.3			4.8	11.6
Domestic/Commercial	0.3					15.8	5.3
Total (excl. Transport)	30.6		13.1			20.6	20.5
10^6 Metric Tons SO_2							
Transformation:							
Power plants	0.32		0.09				0.03
Coke/briquettes	0.08						0.16
Energy sector							
Industry	0.22	0.01				0.03	0.58
Domestic/Commercial	0.01					0.10	0.27
Potential emissions	0.63	0.01	0.09			0.13	1.04
Fuel desulphurisation							

FGD: −0.27 to −0.44   Total Emissions: 1.46 to 1.63

ash during combustion. Sulphur retention in lignite combustion can vary greatly, from as little as 5 per cent for lignites low in alkali metals to as much as 50 per cent for high alkali lignites. This has been taken into account where information was available; otherwise a 5 per cent retention has been assumed.

While the production of coke and lignite briquettes is not a combustion process, emissions from this source have been included since the sulphur released to the coke oven gas is often emitted as SO_2 upon combustion of the gas for underfiring of the coke ovens or elsewhere. During the coking operation approximately one-third of the sulphur in the coal is released to the coke oven gas. When coke is used in the steel-making process the sulphur retained in the coke is not emitted but remains largely in the slag from the steel-making. In some countries, notably Japan and Germany, desulphurisation of coke oven gas is widespread, and emissions from coking operations may be considered negligible.

To estimate the average sulphur content for residual fuel oils in Europe in 1974, the CONCAWE Sulphur Grid Method (9) was applied. Given the crude "package", the method allows the calculation of the sulphur content of the residual fuel oil available for inland consumption. Since the method was developed for a European crude oil and processing situation, it was considered inappropriate for the North American situation. However, there are sufficient similarities between the Japanese and European refinery patterns for the method to be applied for Japan.

JAPAN

Petroleum

The principal point of note from the emissions projections for Japan in Tables 1-3 is that by far the greatest contribution to total SO_2 emissions in Japan will arise from the combustion of residual fuel oils.

The total oil and NGL requirements for energy purposes (including the requirement for electricity generation) for Japan in 1985 compared with 1974 are given in Table 13.

In 1974, imported crude oil represented approximately 90 per cent of all net imports of petroleum substances into Japan. The imports originated from 14 major oil producing countries with Iran, Saudi Arabia and Indonesia being the largest suppliers.(5)

Table 14 gives a broad indication of the levels of sulphur contents of crude oil imported into Japan in 1974.

Table 13

OIL REQUIREMENTS AND IMPORTS IN JAPAN, 1974 AND 1985
(OECD World Energy Outlook (2)

Million metric tons oil equivalent, mtoe

	1974	1985 Reference Case	1985 Accelerated Policy Case
Production	0.7	3.4	3.4
Net imports	263.1	441.5	382.2
Bunkers (-)	-17.9	-34.0	-29.4
Total energy requirements	238.3	410.9	356.2
Electricity generation	-53.0	-56.1	-40.9

Table 14

SULPHUR LEVELS OF CRUDE IMPORTS INTO JAPAN, 1974

Sulphur Level	% Crude Imports, 1974
Low Sulphur (<1.0%)	21
Medium Sulphur (1.0-2.0%)	38
High Sulphur (>2.0%)	41

Using the distillate splits given in Table 15 and the parameters of the Japanese refining situation, application of the CONCAWE Sulphur Grid Method (9) results in an estimated average sulphur content for domestically produced residual fuel oil before any desulphurisation, of 2.2 per cent in 1974. The average sulphur content for imported fuel oil is reported to be much lower at 0.35 per cent.(20) Thus the average sulphur content for the combined fuel oil pool is 2.0 per cent.

Given the 194,000 metric tons of direct hydrodesulphurisation capacity and the 668,000 tons of indirect (vacuum gas oil) capacity installed by the end of 1973,(20) it is estimated that, in 1974, about 790,000 metric tons of sulphur were removed from the residual oil pool. This is equivalent to a reduction of 1.58×10^6 metric tons of SO_2 emissions.

In order to examine the impact of the possible 1985 crude supply patterns on the average sulphur content of residual fuel oils, the CONCAWE Sulphur Grid Method (9) has been applied for two different supply scenarios for each of the Reference and Accelerated Policy

Table 15

AVERAGE YIELDS AND SULPHUR CONTENTS OF CRUDE OILS (%)
(as prevailing in 1973)

	Light Distillate (below 150°C)		Middle Distillate (150-350°C)		Residuum (350°C+)	
	Yield	S cont.	Yield	S cont.	Yield	S cont.
Abu Dhabi	18.4	0.06	40.0	0.73	41.6	2.37
Algeria	21.7	0	43.3	0.06	35.0	0.24
Indonesia	10.0	0	33.4	0.02	56.6	0.22
Iran	14.5	0.10	30.4	0.80	55.1	2.52
Iraq	17.1	0.05	37.6	0.86	45.3	3.76
Libya	15.6	0.01	37.3	0.16	47.1	0.46
Kuwait	13.0	0.04	31.0	0.99	56.0	4.06
Neutral Zone	11.5	0.02	27.5	1.02	61.0	4.33
Nigeria	12.2	0	47.0	0.13	40.8	0.35
Qatar	19.1	0.07	40.1	0.79	40.8	2.80
Saudi Arabia	13.9	0.04	33.9	1.00	52.2	3.25
Venezuela	4.2	0.04	22.8	0.73	73.0	2.50
Australia	21.8	0.02	44.6	0.24	33.6	0.23

cases. To do this, it has been assumed that the refinery pattern will be such that the expected demand for light products can be met.

First, it is worthwhile considering the constraints on crude supplies which will be available to Japanese refineries in 1985. In the last two or three years Japan has increasingly purchased from the Middle East (in the period April-September 1976, 78.6 per cent of crude imports were from that area).

In the period to 1985 it seems that Japan could become increasingly dependent on Middle Eastern crudes, particularly those with higher sulphur contents, like Saudi Arabian crudes. In fact, Japan has, since 1973, been developing closer political and economic ties with Middle Eastern producers, with hopes that this will help secure oil supplies.

At the same time, Japan hopes to reduce (probably only fractionally) the geographical dependence on the Middle East. The likely release of supplies of Indonesian crudes from US West Coast markets, as Alaskan oil comes on stream, may lead to more Indonesian crudes going to Japanese markets. This will depend on relative prices with other crudes. Small quantities of imports may come from increased oil production in the Pacific Area, e.g. from Brunei and Malaysia.

Hence a "best" supply case for 1985, from the point of view of sulphur content, would be that where Japan's crude requirements additional to 1974 include the highest proportion of available low sulphur Indonesian crudes. The crude supply would consist of an additional supply of 50 million metric tons/yr of Indonesian crudes, with the balance of the increased demand being Saudi Arabian crudes. From this scenario, average residual fuel oil sulphur contents of 2.4 per cent in the Reference Case or 1.5 per cent in the Accelerated Policy Case, both before hydrodesulphurisation, could be expected.

For a "worst" supply scenario where the additional crude requirement is imported wholly from the Middle East, the corresponding average sulphur content is 2.8 per cent in both the Reference Case and the Accelerated Policy Case.

Table 16 summarises the potential SO_2 generation from combustion of the fuel oils, the removal by the 1974 hydrodesulphurisation (HDS) capacity and the removal by the HDS capacity which can be projected for 1985. If, in Table 16, it is assumed that the 1985 flue gas desulphurisation (FGD) capacity forecast in Chapter III is fitted predominantly (92 per cent) to oil fired boilers, then the resultant SO_2 recovery by the combined 1985 HDS and FGD capacities and the resultant emissions will be as shown. For comparison, a Japanese Government Report (10) forecast that the total SO_2 emissions reduction through HDS and FGD would be 6.7 million metric tons in 1985.

Table 16

POTENTIAL SO_2 GENERATION FROM FUEL OIL COMBUSTION
AND REMOVAL BY HDS AND FGD IN JAPAN, 1974 AND 1985

10^6 metric tons SO_2

	1974	1985 Reference Case		1985 Accelerated Policy Case	
		"Worst"	"Best"	"Worst"	"Best"
Potential Generation	3.82	8.97	7.68	7.68	4.11
Removal by 1974 HDS	-1.58				
Removal by Projected 1985 HDS		-4.40	-4.20	-3.40	-1.80
Projected Removal by 1985 HDS and FGD		-6.05	-5.85	-5.05	-3.10
Emissions	2.24	2.92	1.83	2.63	1.01

All gas oil in Japan is desulphurised. In fact, in 1975 the gas oil desulphurisation capacity was 59,740 metric tons of gas oil compared with a production of 29,086 metric tons. Consequently, SO_2 emissions from this source have been estimated at 0.19 million metric tons in the 1985 Accelerated Policy Case, and 0.23 million metric tons in the 1985 Reference Case, compared with 0.10 million metric tons in 1974. This corresponds to 0.3 per cent average sulphur content.

Finally, it has been assumed that direct burning of crude petroleum will continue to take place in Japan in 1985. In 1974 the direct use of 20 million metric tons of crude oil in power stations was responsible for approximately 0.240 million metric tons of SO_2 emitted. The burning of low sulphur crudes (about 0.6 per cent S) is a preferred strategy of the electric power industry for reducing SO_2 emissions below those which would result from burning an equivalent quantity of residual fuel oil. This practice also provides a lower nitrogen content fuel with the benefit of lower nitrogen oxides emissions. Although a greater economic return might be possible from transforming this crude into saleable products, this would need to be measured against the possible increased cost of SO_x and NO_x control if residual fuel oil replaced the crude burned directly.

Coal

Table 17 gives details of solid fuel production and trade levels for Japan in 1974 and estimates for 1985, for both the Reference and Accelerated Policy Cases. Substantially increased coal imports will be required in Japan in 1985.

Of Japan's 1974 imports of coal, all were coking coal, mostly North American (25.4 million metric tons of coking coal from the US and 9.8 million metric tons from Canada) or Australian (20.9 million metric tons). All steaming coals were indigenously produced. Due to the nature of the demand, most of the steaming coal in trade in 1985 is likely to have high heating values and relatively low sulphur contents. Certainly, the Australian coals to be exported to the Japanese market are of very low sulphur content (less than 1 per cent), and the North American coals exported could also be of low sulphur content.

Table 17

SOLID FUEL PRODUCTION AND TRADE IN JAPAN - 1974 AND 1985
(OECD World Energy Outlook (2))

Million metric tons oil equivalent

	1974	1985 Reference Case	1985 Accelerated Policy Case
Production	15.6	12.7	12.7
Net Imports	47.3	75.4	75.4
Total Energy Requirements	63.1	88.1	88.1
Electricity Generation	-8.5	-14.4	-14.4

For combustion of coals in both 1974 and 1985 the average sulphur content is estimated at 0.7 per cent weight (Report of the Joint Ad Hoc Group).(7) The sulphur content of coals combusted in Japan in

1985 might be less if all of increased imports of low sulphur coal from Australia are for steaming. However, this small change in coal sulphur content would not be expected to have a significant effect on the estimate of total SO_2 emissions.

As the removal of sulphur from coke-oven gas and other gases is widespread in Japan it has been assumed that there would be no SO_2 emissions from secondary transformation of coal in 1974 or 1985.

The potential SO_2 emissions from coal combustion and transformation to secondary energy sources in Japan are expected to grow from 0.16×10^6 metric tons in 1974 to 0.31×10^6 metric tons in either 1985 case, before any allowance for SO_2 recovery by flue gas desulphurisation.

In conclusion, Japan's own production (1 per cent to 3 per cent sulphur content) will decline over the period 1976 to 1985. Future supplies of coking coal are expected from Australia and Canada, as Japan is apparently anxious to diversify its supplies from increasingly expensive US sources. As demands for steaming coals are likely to be met from Australia, the indications are that there may be marginal reductions in the average sulphur content of coal burnt over the period under review.

OECD EUROPE

Petroleum

It can be seen from Tables 4 to 6 that in Europe, unlike Japan, SO_2 emissions from both coal and residual fuel oil combustion make a significant contribution to the total SO_2 emissions. The total European oil requirements for energy purposes and the requirement for electricity generation are compared for 1974 and 1985 in Table 18.(2)

Table 18

OIL REQUIREMENTS AND IMPORTS IN OECD EUROPE - 1974 AND 1985
(OECD World Energy Outlook (2))

mtoe

	1974	1985 Reference Case	1985 Accelerated Policy Case
Production	22.3	221.6	221.6
Net Imports	708.1	738.0	554.2
Bunkers (-)	-36.4	-55.4	-40.6
Total Energy Required	667.5	904.2	735.2
Electricity Generation	-83.8	-160.0	-108.1

97.5 per cent of petroleum substances imported (net) into the OECD Europe area in 1974 were in the form of crude oils. The biggest suppliers in 1974 were Saudi Arabia (30.6 per cent) and Iran (17.2 per cent).(5) In 1974, 45 per cent of the crude oil imports had relatively high sulphur content, 30 per cent were of medium sulphur content and 25 per cent were low sulphur.

Given the average yields and sulphur contents of Table 15 for Japan, the CONCAWE Sulphur Grid Method (9) has been used to estimate the average residual fuel oil sulphur content in Europe in 1974, before any desulphurisation, to be 2.5 per cent. Historical information received from oil industry sources confirms this estimate. This resulted in the potential generation of SO_2 from combustion of residual fuel oils in 1974 of 11.21×10^6 metric tons. Table 4 shows that there was no reduction in these emissions by hydrodesulphurisation of heavy fuel oils, although it is known that the installed desulphurisation capacity was $6,600 \times 10^3$ metric tons by the indirect process and 425×10^3 metric tons by the direct process at the end of 1975. However, the degree to which the capacity has been utilised is not known.

In examining the crude oil requirement for Europe in 1985, it is of interest that Europe is the one OECD region for which net oil imports are not expected to increase greatly between 1974 and 1985. The main reasons for this are rapid increases in production of North Sea oil to 195 mtoe over the forecast period. These crudes are relatively light and relatively low in sulphur (e.g. Beryl 0.36 per cent, Ekofisk 0.21 per cent and Forties 0.28 per cent). If consumed in Europe, they should therefore facilitate the reduction of sulphur emissions.

On the basis of the existing refining pattern it appears unlikely that European demand for light crudes will be as great as the estimated supplies of North Sea crudes in 1985. Whether or not supplies of North Sea crudes supplement or displace present imports of light crudes, and to what degree, clearly depends upon the production and export policies that will be adopted by the producers, Norway and the United Kingdom. It is possible that some UK produced crude would be exported to the United States, and that prices established for North African light crudes will make them competitive with North Sea crudes in Europe.

The consequent "best" crude supply scenario, from the point of view of sulphur content, would be for all North Sea crudes (200 million metric tons) to be consumed in Europe in addition to maintaining the present level of imports of other light, low-sulphur crudes. For the Accelerated Policy Case, since the imports are reduced from 1974, it is expected that, in the "best" scenario, the medium sulphur crude imports would be partly displaced by the North

Sea crudes. In the Reference Case additional imported crude requirements over 1974 would probably come from the high sulphur Saudi Arabian crudes to meet the demand of 960 million metric tons of oil. Under these assumptions the average residual fuel oil sulphur contents, before any desulphurisation, of 2.1 per cent in the Reference Case and 1.7 per cent in the Accelerated Policy Case have been estimated.

If the North Sea crudes are not totally used in Europe, or merely displace present imports of light low sulphur crudes, and high sulphur Middle Eastern crude oil is used to meet additional import requirements, then a "worsening" of the sulphur emissions from combustion of heavy fuel oil could result. The range of average sulphur contents of heavy fuel oil would be from 2.7 per cent in the Reference Case to 2.3 per cent in the Accelerated Policy Case.

Table 19 summarises the impact of the "best" and "worst" 1985 crude supply cases.

Table 19

POTENTIAL SO_2 GENERATION FROM FUEL OIL COMBUSTION IN EUROPE, 1974 AND 1985

10^6 metric tons SO_2

	1974	1985 Reference Case		1985 Accelerated Policy Case	
		"Worst"	"Best"	"Worst"	"Best"
Potential generation	11.20	15.11	11.74	10.48	7.73

It is apparent from Table 4 that SO_2 emissions from combustion of gas oils in Europe are far more significant than in Japan, contributing 1.35 million metric tons of SO_2 in 1974. This estimate is based upon a gas oil sulphur content of 0.5 per cent in 1974, while, for 1985 it has been assumed that EEC regulations limiting gas oil sulphur content to 0.3 per cent would be met not only in EEC countries but also throughout the rest of OECD Europe. For both 1985 forecasts, these regulations produce an improvement over the position in 1974.

In conclusion, Table 19 clearly shows that there is potential for reducing the sulphur content of residual fuel oils if the North Sea production and European import policies are such that the refining and consumption of North Sea crude production are within Europe and there are supplemental imports of other low sulphur crudes.

Coal

Solid fuel production and imports for OECD Europe in 1974 and forecasts for 1985 are shown in Table 20. While production of coal in OECD Europe declined from 334 Mtoe in 1960 to 206 Mtoe in 1974, the projections for both the Reference and Accelerated Policy Cases suggest that this decline will be halted and that a small increase in production to 1985 might be achieved. UK production of hard coals should increase slightly while Spanish and Turkish production of lignites should grow rapidly; French production is expected to be halved by 1985 and German production to register a small decline. These trends predict a somewhat higher 1985 market share for lignite than at present. The 1985 coal consumption forecasts also imply substantially increased coal imports into Europe.

Table 20

SOLID FUEL PRODUCTION AND TRADE IN OECD EUROPE 1974 AND 1985
(OECD World Energy Outlook (2))

mtoe

	1974	1985 Reference Case	1985 Accelerated Policy Case
Production	205.8	215.6	215.6
Net Imports	35.5	56.0	75.2
Total Energy Requirements	253.7	271.6	290.8
Electricity Generation	-115.4	-133.8	-144.8

Unlike emissions from fuel oil and gas/diesel oil combustion which were estimated on a regional basis, the SO_2 emissions from coal combustion have been calculated country by country. This procedure was necessitated by the complex pattern of coal trade and the variations of sulphur contents from one country to another. Only the total emissions for OECD Europe from this source have been tabulated in this report (Tables 4 to 6). The sulphur contents of coals combusted in OECD Europe in 1974 are summarised in Table 21.

Without allowing for the planned cleaning of German hard coals to an average sulphur level of 1 per cent, the average sulphur level of hard coals for combustion in OECD Europe is expected to increase slightly by 1985 partly as a result of the increased United Kingdom production of coals with above average sulphur content. Table 22 summarises the information in Tables 5 and 6, showing both the potential SO_2 emissions and the expected reduction in emissions from the cleaning of German coals to 1 per cent sulphur. In the 1985 Accelerated Policy Case the coal import requirements are higher than

Table 21

AVERAGE SULPHUR CONTENT OF COALS(*) FOR COMBUSTION
IN OECD EUROPE, 1974

% weight sulphur

Country	% Sulphur
Austria	1.2
Belgium	1.0
Denmark	0.8 for power plants
	0.95 for other sectors
Finland	1.0
France	0.7
Germany: hard coal	1.23
lignite	0.5
Greece	1.0
Ireland	1.3
Italy: hard coal	0.9
lignite	0.8
Luxembourg	1.0
Netherlands	1.0
Norway	1.2
Portugal	1.2
Spain: hard coal	1.2
lignite	5.0
Sweden	1.0
Switzerland	1.2
Turkey	2.0
United Kingdom	1.46 (varies with sector)

*) Lignites have been assumed to have the same sulphur content as hard coals unless stated otherwise.

in the Reference Case and the additional amount is likely to come largely from the United States (about 28 million metric tons) with a higher sulphur content (1.8-1.9 per cent). Consequently, the average sulphur content of coals combusted would be higher in that case than in the Reference Case.

The SO_2 emission from lignites mined in Europe will increase from about 1.0 per cent w SO_2 emitted in 1974 to about 1.5 per cent emitted in 1985, based on the Secretariat's estimate of lignite production in 1985 and the sulphur levels for lignites in 1974. Major contributors to this increase are the increased production of Spanish and Turkish lignites. Spanish lignites, in particular, can be of very high sulphur content (up to 7 per cent by weight).

In summary, the substantial increase in coal use in OECD Europe by 1985 could lead to a worsening of emissions arising from high sulphur lignites as their share of the market increases and the possible imports of higher sulphur United States steaming coals. The further coal cleaning planned in the Federal Republic of Germany will only marginally offset the increased emissions for the OECD Europe region as a whole.

Table 22

SO_2 GENERATION FROM SOLID FUEL COMBUSTION AND TRANSFORMATION TO SECONDARY ENERGY SOURCES FOR OECD EUROPE, 1974 AND 1985

10^6 metric tons SO_2

	1974	1985 Reference Case	1985 Accelerated Policy Case
From combustion	6.71	9.00	9.87
From secondary transformation	0.46	0.54	0.54
Removal by additional coal cleaning		-0.20	-0.20
Total	7.17	9.34	10.21

NORTH AMERICA

Petroleum

Total North American requirements for oil in 1974 and 1985 are presented in Table 23 together with import requirements.

Indigenous production in Canada in 1974 was 82.4 mtoe and this is expected to decline to 55 mtoe in 1985. Although frontier production may start during the 1980s, none is included in the estimates for 1985, but tar sands are expected to produce 13.5 mtoe.

Table 23

OIL REQUIREMENTS AND IMPORTS IN NORTH AMERICA, 1974 AND 1985 (2)

mtoe

	1974	1985 Reference Case	1985 Accelerated Policy Case
Production	592.1	650.9	771.4
Net Imports	281.0	531.1	246.1
Marine Bunkers	-20.1	-31.1	-14.7
Total Energy Requirements	844.7	1,150.9	1,002.8
Electricity Generation	-84.0	-87.4	-56.6

In the United States, Alaskan oil is expected to contribute 125 mtoe in the 1985 Reference Case and 150 mtoe in the Accelerated Policy Case. Production from OCS, in particular National Petroleum

Reserve 1, should total 15 mtoe in 1985. Production from the Beaufort Sea of 20 mtoe is assumed in the Accelerated Policy Case. Given these estimates production from existing United States fields is likely to be about 414 mtoe in the 1985 Reference Case, and about 505 mtoe in the Accelerated Policy Case (compared with 498.3 mtoe in 1974).

In 1974, of the net 281 million metric tons oil equivalent of crude and products imported into North America, about 40 per cent (112.6 million metric tons) was in product form, in large part residual fuel oil from **Caribbean** refineries.

From statistics of fuel oil deliveries to United States utilities,(11) it is known that the average sulphur content of heavy fuel oil delivered in 1974 was 1.1 per cent. In Table 7 this percentage is used. It was observed that utilities burned fuel oil that was, on average, lower in sulphur than that consumed in other sectors, for which an average sulphur content of 1.5 per cent has been used. The National Petroleum Council reported that the 1971 average sulphur level of domestically refined residual fuel oil was 1.4 per cent and for imports was 1.5 per cent.(12) The emission estimates in Table 7 for residual combustion compare reasonably well with the US EPA 1973 estimates (13) of 3.6 million metric tons SO_2, considering the 7 per cent decrease in residual oil consumption from 1973 to 1974.(3)

For Canada, in preparing Table 10, it was noted that the average sulphur content of residual fuel oils consumed in 1968 was 2.5 per cent.(7) An examination of the pattern of imports of crude oils and residual fuel oil product between 1968 and 1974 indicates little change in that sulphur content. Unfortunately, because of the differences between the European and North American crude oil and refinery patterns, the CONCAWE sulphur grid method could not be used to give a more accurate assessment.

Critical to an estimation of the sulphur levels of North American residual fuel oils in 1985 is the balance between low sulphur indigenous crude production and higher sulphur imported crudes. Major uncertainties surround the utilisation of Alaskan production in North America. Although political factors seem to **dictate** that this indigenous production will be consumed in the United States, the market identified to absorb it - the West Coast - has a limited capacity. This implies that between 0.8 m b/d and 1.3 m b/d of Alaskan crude which cannot be absorbed on the West Coast may be shipped from Port Valdes, through the Panama Canal to the East Coast refineries, pipelined to refineries in states bordering Canada which are losing their crude supplies, pipelined to Texas, though this proposal has encountered substantial environmental opposition, or finally exported, probably to Japan.

In addition to these domestic uncertainties, others exist in the international oil trade. Venezuela faces declining production; currently a large proportion of the USA's and Canada's imports of crudes and products originate there. But on the other hand, Mexico is developing into a major new oil exporter, with anticipated exports of medium sulphur crudes. There could also be some exports of North Sea crudes to the United States in the 1980s (the North Sea is closer to the United States than the West or North African oil producers), or alternatively exports to the United States of the light and medium African crudes which will be displaced from European markets by North Sea crudes. Given these uncertainties, no attempt can be made to forecast the change, if any, in the sulphur levels of residual fuel oil consumed in North America in 1985 and the levels prevailing in 1974 have been used in the emissions forecasts.

Regarding the sulphur content of gas oil, the large gas oil desulphurisation capacity would appear to indicate that 0.3 per cent will be the average for both 1974 and 1985. The US EPA (13) appeared to use this value in their estimate of emissions from gas oil combustion in 1973 of 0.59×10^6 metric tons SO_2.

Although there are plans to install some capacity for hydrodesulphurisation of residual fuel oils in North America, the potential sulphur removal of this capacity cannot be determined without a forecast of the changing crude supply patterns.

Coal

It is clear that by far the greatest contribution to SO_2 emissions from fuel combustion in North America comes from coal. This is a result not only of the high consumption but also of the relatively high average sulphur content of indigenous coal reserves. With the acceleration of coal production to 1985 leading to a higher share of coal in total energy production, potential emissions from this fuel could increase substantially.

Coal production and exports for the United States in 1974 and estimates for 1985 are shown in Table 24.

Table 24 shows the expected substantial increase in production, much of which will be from the Western coalfields. These Western coals will be a mixture of bituminous, sub-bituminous and lignites with a lower than average heating value, but also a lower sulphur content.

Table 25 shows the sulphur contents of United States coals for combustion in 1974, estimated for each production region,(14) (15) and (16). The sulphur content for Western region raw coal is the average for hard coal and lignite. Only some of the hard coals from the Western region would be subjected to cleaning.

Table 24

SOLID FUEL PRODUCTION AND TRADE IN THE UNITED STATES,
1974 AND 1985

mtoe

	1974	1985 Reference Case	1985 Accelerated Policy Case
Production	356.8	532.5	532.5
Net Imports	-32.2	-39.9	-55.9
Total Energy Requirement	327.6	492.6	476.6
Electricity Generation	-231.9	-368.0	-355.5

Table 25

1974-75 ESTIMATED SULPHUR CONTENT OF US COALS

Average Sulphur Content, wt. per cent

Production Region	Raw Coal (14)	Cleaned Coal (14)	Coking Coal (100% cleaned) (15)	Combustion Coal (27% cleaned)	Power Plant Coal (16)
Northern Appalachian	3.01	2.06	0.9-1.2	3.2	2.8
Southern Appalachian	1.08	0.97	0.9	1.1	1.4
Eastern Midwest	3.92	2.73	0.8-1.0	3.2	3.5
Western Midwest	5.25	3.91	0.9-1.0	5.4	4.8
Western	0.68	0.54	0.7	0.7	0.5
All Regions	2.56	1.87	0.9	2.3	2.2

Similar estimates have been made for the sulphur content of United States coals in 1985 based on a forecast of production for each region by the US Federal Energy Administration.(17) The total amount of combustion coal projected to be cleaned is approximately 180 million metric tons (or 23 per cent of the combustion coal). The resulting average sulphur content for combustion coal is 1.64 per cent in the Reference Case and 1.59 per cent in the Accelerated Policy Case. The lower sulphur content in the Accelerated Policy Case is a consequence of additional exports to Europe of 28 million metric tons of steaming coal from Eastern regions with higher than average sulphur content (1.8-1.9 per cent).

Canadian coal production and imports in 1974 and 1985 are shown in Table 26.

In 1974 Canada exported only coking coals, mainly from its Western provinces, and imported both steaming and coking coals from the United States into its Eastern provinces. Only minor amounts of

Table 26

SOLID FUEL PRODUCTION AND TRADE IN CANADA, 1974 AND 1985

mtoe

	1974	1985 Reference Case	1985 Accelerated Policy Case
Production	12.5	29.4	29.1
Net Imports	1.4	-5.3	-5.3
Total Energy Requirement	14.9	24.1	23.8
Electricity Generation	-9.0	-16.4	-16.4

combustion coals were cleaned. Knowing the sulphur contents of Canadian coals (Table 27) and the fact that the bulk of the low sulphur Alberta sub-bituminous (hard) coal production is consumed in power stations, the sulphur contents of coals consumed in each sector have been estimated to be: for hard coal, 1.4 per cent for power stations and 2.2 per cent for other combustion; and for lignite, 0.75 per cent.

Table 27

ESTIMATED SULPHUR CONTENT OF CANADIAN COALS

% weight sulphur

Production Region	Mined Coal (Some cleaned)	Coking Coal	Combustion Coal
Nova Scotia	2.3	1.1	3.0
New Brunswick	5.0		5.0
Alberta: bituminous	0.4	0.4	0.4
sub-bituminous	0.2		0.2
British Columbia	0.3	0.3	0.3
Saskatchewan: Lignite	0.75		0.75
Imports from US: 1974			2.1
1985			2.7

SO_2 emissions from coal utilisation in North America in 1974 and 1985 are summarised in Table 28. These emissions have taken into account the sulphur removal by physical cleaning existing in 1974 and projected for 1985 in the United States, which amounts to 2.4 million metric tons and 2.8 million metric tons sulphur respectively. It was not possible to divide these reductions among the coal used for combustion, for coking and for export and to obtain the actual emission reduction.

Table 28

SO_2 EMISSIONS FROM COAL COMBUSTION IN NORTH AMERICA,
1974 AND 1985

10^6 metric tons SO_2

	1974	1985 Reference Case	1985 Accelerated Policy Case
From hard coal	19.08	25.19	23.63
From lignite	0.20	0.40	0.38
Total	19.42	25.59	24.01

In conclusion, the development of Western coal reserves in North America will lead to a significantly reduced average sulphur content of coals for combustion there by 1985. If even greater emphasis is placed on coal cleaning then there is scope for a further substantial reduction in potential SO_2 emissions.

Chapter III

POTENTIAL FOR REDUCTION OF SO_2 EMISSIONS

Before examining the potential for reducing SO_2 emissions from fuel combustion by the further application of fuel cleaning technologies or post-combustion stack gas cleaning, it is worthwhile reviewing the potential that exists for increasing the supply of naturally low-sulphur fuels.

SUPPLY OF NATURALLY LOW-SULPHUR FUELS

Coal

Coal combustion is a major contributor to SO_2 emissions in both OECD Europe and North America. Of these regions, only North America has significant low sulphur coal resources which could be developed by 1985. This development has been assumed in the emissions forecast of Chapter II. However, should coal production fall behind target the shortfall could be expected in this new development, and the full potential for reducing emissions by supplying naturally low-sulphur coal would not be realised.

OECD Europe, on the other hand, appears to be exhausting much of its low-sulphur reserves. The forecast is for increased production (135 million metric tons per annum) of slightly higher sulphur United Kingdom coals, coupled with a decline in production of low sulphur coals such as French coals. Increased development of lignite reserves in Spain and Turkey will also contribute to higher sulphur levels.

In 1974 most coal in international trade was coking coal. The expanding trade in steaming coals might be thought to indicate potential for improving sulphur levels through importing low sulphur coals. While this may be possible for individual countries dependent upon coal imports, the potential for the OECD European region as a whole is limited. Additional imports of steaming coal from the United States and Canadian East coasts would generally be of higher than average European sulphur content. Quantities of low sulphur steaming coals which will be available from Australia and possibly also India,

Poland and, to a certain extent, South Africa would, in the period to 1985, be small in comparison to total coal consumption.

Petroleum

The potential for reducing SO_2 emissions from residual fuel oil combustion by increasing the availability of low sulphur crudes to refineries has been examined in the preceding Chapter.

It appears that Japan may be able to secure supplies of low sulphur Indonesian crudes (40 to 65 mtoe per annum) which could be released from US West Coast markets as Alaskan crudes come onstream. Alternatively, some low-medium sulphur Alaskan crudes may be available for export to Japan but the necessary permit has not been granted. However, the remainder of the increase in Japanese crude oil demand will most likely come from Middle Eastern producers with whom Japan has developed close economic ties. If Japan is to derive a significant reduction in crude sulphur levels from the purchase of Indonesian or other low sulphur crudes it will also need to minimise purchases of higher sulphur crudes. This entails holding back demand as in the Accelerated Policy Case. In that case, increased low sulphur crudes could bring about a reduction in the average sulphur content of residual fuel oil, before desulphurisation, to 1.5 per cent weight sulphur from the 1974 level of 2.0 per cent. Should the United States compete for these supplies then the Japanese position will not be as good and there may even be an increase on the present crude sulphur content.

As North Sea crude production increases, the potential for a reduction in emissions from fuel oil combustion in OECD Europe appears substantial for the region as a whole. It may be that all the anticipated North Sea production of 200 mtoe in 1985 is available for European consumption in addition to the present level of imports of low sulphur crudes, principally North African. If so, the sulphur content of residual fuel oil available for inland consumption may improve to 2.1 per cent in the Reference Case or 1.7 per cent in the Accelerated Policy Case from 2.5 per cent in 1974. Should the North Sea supplies displace imports of other low sulphur crudes, the improvement in sulphur level will not be as great. Another factor to consider is that already North Sea crudes are being exported to the United States. If the United States purchases a greater proportion of North Sea production in the future and also competes for supplies of low sulphur North African crudes, then the potential to reduce emissions will again be less.

The potential for achieving lower crude oil sulphur levels in North America would not appear to be great. Alaskan crude now coming on stream has an average sulphur content of about 1 per cent which, in fact, is a little higher than much current United States production.

At the same time, this crude may displace rather than supplement some of the lighter imported Indonesian crudes, especially on the Western seaboard. The United States may secure supplies of North Sea oil, or, alternatively, may purchase low sulphur North African crudes released from European markets. This would bring about some reduction in the sulphur content of crudes processed in the United States in 1985. If, as anticipated, medium sulphur Mexican crudes play a more important role in the United States petroleum market, then the overall decline in sulphur levels will be only marginal.

What does emerge from the above discussion is the interdependence of crude oil supplies to Japan, North America and OECD Europe. There will be an increase in the supply of low sulphur crudes available, but much of the increased oil demand will still have to be met from high sulphur crudes. Thus, there will be competition for available low sulphur crudes and not all the OECD regions can expect to achieve substantial reductions in sulphur levels of their crude oil supplies.

COAL CLEANING - NORTH AMERICA

For reasons of data availability, this discussion is limited to an assessment of present coal cleaning practice in the United States and the potential for further cleaning of US coals.

It is unlikely that new coal preparation technology will make any significant breakthrough in the desulphurisation potential of coal from now to 1985. Although novel methods currently being investigated, such as magnetic separation and chemical leaching processes, may be developed satsifactorily for commercial application prior to 1985, quantitative contributions of these methods in coal desulphurisation are expected to be insignificant up to 1985.

The types of coal cleaning equipment used in the United States are shown in Table 29. Because of environmental pressures and the resulting increased need for sulphur reduction, it is anticipated that the majority of new coal cleaning plants built will employ multi-stage operations including dense-medium processes, concentrating tables and froth flotation. Hence, it is assumed that the amount of coal handled by jigs, pneumatic methods, launders and classifiers will remain at the present level, but that the amount handled by dense-medium processes, concentrating tables and froth flotation will increase substantially and approximately account for the estimated increase in total coal to be cleaned in 1985.

The quantity of coal cleaned in the United States in 1977 and rough estimates of the quantity of sulphur removed are summarised in Table 29. These numbers include combustion coals, coking coals and exports. The estimates are only rough approximations because the data are based on a regional sampling of US coals analysed by the

US Bureau of Mines. The coal production in the United States is broken down by regions in the following manner:

1. Northern Appalachian Region (Pennsylvania, Ohio, Maryland and West Virginia (North)) - contains hards coals with 0.5 to 5 per cent sulphur;
2. Southern Appalachian Region (Virginia, Alabama, West Virginia (South), Kentucky (East) and Tennessee) - contains hard coals with 0.3 to 2 per cent sulphur;
3. Eastern Midwest Region (Illinois, Indiana and Kentucky (West)) - contains hard coals with 2 to 5 per cent sulphur;
4. Western Midwest Region (Arkansas, Iowa, Kansas, Missouri and Oklahoma) - contains hard coals with 3 to 6 per cent sulphur; and
5. Western Region (Arizona, Alaska, Colorado, Montana, New Mexico, North Dakota, Texas, Utah, Washington and Wyoming) - contains hard coals and lignites with 0 to 1 per cent sulphur.

Table 29

TYPE OF EQUIPMENT USED IN CLEANING OF BITUMINOUS COAL

Type of Equipment	1974 (18)		1985 Estimates (17)	
	Amount, 10^3 short tons	Weight, %	Amount, 10^6 short tons	Weight, %
Jigs	129,302	48.8	130.0	34.8
Dense-Medium Processes	82,283	31.0	156.6	41.4
Concentrating Tables	28,869	10.9	54.4	14.6
Flotation	10,863	4.1	20.4	5.5
Pneumatic Methods	7,557	2.8	7.6	2.0
Launders and Classifiers	6,275	2.4	6.2	1.7
Total	265,150	100.0	373.2	100.0

The present coal preparation practices separate the raw coal into a clean coal and a waste. In operating the units, there is a trade-off between the sulphur removal and the loss of coal into the waste. This latter loss is referred to as percentage Btu recovery. In Table 30, the desulphurisation potential is estimated at two levels of coal preparation:

Level A - Coal is crushed to 1-½ inch top size and beneficiated with 90 per cent Btu recovery;

<u>Level B</u> - Coal is crushed to 1-½ inch top size and beneficiated with 80 per cent Btu recovery.

It should be noted that the estimated sulphur contents of cleaned coal are those obtained by float-sink analysis at the given Btu recoveries, and they may differ from the actual values obtained in commercial coal cleaning plants. There are two major factors which affect the sulphur removal by coal cleaning:

a) Inefficiency of coal cleaning equipment: the sulphur content of commercially cleaned coal is generally higher than that obtained by float-sink analysis if the Btu recoveries are the same.

b) Btu recovery: In general, for any given coal, the lower the Btu recovery, the lower the sulphur content of the cleaned coal. The average Btu recovery in commercial coal cleaning plants is approximately 80 per cent.

Considering these two factors, the sulphur contents of cleaned coal by Level A (90 per cent Btu recovery) shown in Table 30 appear to be reasonable values for the purpose of estimating sulphur removal for current coal cleaning practice in the United States. On the other hand, the sulphur contents of coal cleaned by Level B are the lowest possible values which could be obtained by cleaning the coal with 100 per cent efficiency and an 80 per cent Btu recovery. The quantity of sulphur removed in 1974 is therefore estimated to be 2.7 x 10^6 short

Table 30

MECHANICAL CLEANING OF US BITUMINOUS COAL AND LIGNITE IN 1974

Region	Total Coal Production 10^6 Short Tons	Mechanical Cleaning 10^6 Short Tons (18) Cleaned Coal	Total Sulphur Content wt. per cent (14)			Estimated Sulphur Removal 1000 Short Tons	
			Raw Coal	Level A Cleaned	Level B Cleaned	Level A Cleaned	Level B Cleaned
Northern Appalachian	179.4	86.6	3.01	2.06	1.83	1,112	1,673
Southern Appalachian	198.3	82.0	1.08	0.97	0.96	189	320
Eastern Midwest	133.8	83.7	3.92	2.73	2.57	1,361	1,951
Western Midwest	8.7	0.8	5.25	3.91	3.76	16	24
Western	83.2	12.0	0.68	0.54	0.51	26	41
Total	603.4	265.1	2.56	1.87	1.74	2,704	4,009

tons. The potential sulphur removal with the existing facilities would be 4.0×10^6 short tons, but with the subsequent energy penalty of about 11 per cent due to coal loss to the waste.

The coal used for coking and export in 1974 was 149.3×10^6 short tons (averaging 1.0 per cent sulphur), all of which is assumed to have been mechanically cleaned. Hence, only 27 per cent of the coal used for combustion, or 115.8×10^6 short tons, is mechanically cleaned. This indicates an average sulphur content for all combustion coals of 2.3 per cent. The Level A cleaning was assumed for the emission survey of Chapter II and in Table 7.

The Quantity of US Coal which would be Desulphurised in 1985

A forecast of the quantity of US coal to be cleaned and estimates of the quantity of sulphur which could be removed in 1985 are summarised in Table 31. In addition to Levels A and B, previously defined, the following level of coal preparation is included in Table 31:

Level C - Coal is crushed to 3/8-inch top size and beneficiated with 80 per cent Btu recovery.

Table 31 shows that the total coal to be produced in 1985 is estimated to be 1040×10^6 short tons of which an estimated 373.2×10^6 short tons will be mechanically cleaned. The quantity of sulphur removed is estimated to be in the range 3.2×10^6 to 5.0×10^6 short tons.

The coal to be used for coking and export in 1985 is estimated to be 182.8×10^6 short tons for the Reference Case (averaging 1.0 per cent sulphur). Hence, mechanically clean coal used for combustion is estimated to be 190.4×10^6 short tons with the average sulphur content for all combustion coals in the range 1.57 to 1.64 per cent.

For the Accelerated Policy Case, the average sulphur content of coals for combustion domestically would be in the range 1.57 to 1.59 per cent.

Although the above estimates of the quantity of coal to be cleaned in 1985 for combustion purposes are likely to be conservative, the most conservative Level A cleaning was assumed in forecasting the 1985 emissions for the United States.

In order to determine the maximum possible sulphur removal, it may be useful to examine the application of mechanical cleaning to as much coal as is feasible. The quantity of sulphur removed would be augmented considerably if all high sulphur Northern Appalachian, Eastern Midwest and Western Midwest coals were to be cleaned and if the same proportion of low sulphur Southern Appalachian and Western coals as at present were to be cleaned. At Level A cleaning, the estimated sulphur removal would be increased to 5.56×10^6 short tons from the 3.16×10^6 short tons shown in Table 31. For Levels B and C,

Table 31

ESTIMATED MECHANICAL CLEANING OF BITUMINOUS COAL AND LIGNITE IN 1985

Region	Total Coal Production[17] (10^6 Short t.)	Estimated Mechanical Cleaning[17] (10^6 Short t.) Cleaned Coal	Total Sulphur Content, wt. per cent[14]				Estimated Sulphur Removal (10^6 Short t.)		
			Raw Coal	Level A Cleaned	Level B Cleaned	Level C Cleaned	Level A Cleaned	Level B Cleaned	Level C Cleaned
Northern Appalachian	183	88.3	3.01	2.06	1.83	1.61	1.13	1.71	1.90
Southern Appalachian	322	133.2	1.08	0.97	0.96	0.94	0.31	0.52	0.55
Eastern Midwest	156	97.6	3.92	2.73	2.57	2.47	1.59	2.27	2.37
Western Midwest	9	0.9	5.25	3.91	3.76	3.45	0.02	0.03	0.03
Western	370	53.2	0.68	0.54	0.51	0.53	0.11	0.18	0.17
Total	1,040	373.2	2.20	1.63	1.53	1.45	3.16	4.71	5.02

Table 32

POTENTIAL SULPHUR REDUCTION OF US COALS IN 1985 AT 1.3 AND 1.6 SPECIFIC GRAVITY SEPARATION

Region	Total Coal Production (10⁶ short tons)	Clean Product 1.3 SG Float (10⁶ short tons)	Middlings 1.3 SG Sink 1.6 SG Float (10⁶ short tons)	Total Sulphur Content (wt. %)(14)		
				Raw Coal	Product	Middlings
Northern Appalachian	183	69.8	113.2	3.01	1.41	2.56
Southern Appalachian	322	185.1	136.9	1.08	0.86	1.10
Eastern Midwest	156	84.5	71.5	3.92	2.35	3.53
Western Midwest	9	4.2	4.8	5.25	2.93	4.74
Western	370	217.1	152.9	0.68	0.55	0.60
Total	1,040	560.7	479.3		1.05	1.68

Table 33

TWO PRODUCT COAL CLEANING STRATEGY FOR THE 1985 REFERENCE CASE

Region	Combustion Coal Consumed Domestically	Clean Product		Product for FGD		Total Sulphur Content (wt. %)	
		Part Cleaned** Combustion Coal	Cleaned* Combustion Coal	Middlings* Combustion Coal		Clean Product	Product for FGD
	(10^6 short tons)	(10^6 short tons)	(10^6 short tons)	(10^6 short tons)			
Northern Appl.*	151	–	37.8	113.2		1.72	2.56
Southern Appl.**	200	200	–	–		1.11	–
Eastern Midwest*	145.3	–	73.8	71.5		2.47	3.53
Western Midwest*	8	–	3.2	4.8		3.53	4.74
Western**	352.9	352.9	–	–		0.66	–
Total	857.2	552.9	114.8	189.5		1.07	2.98

* Cleaning of 100 per cent of production at 1.3 and 1.6 specific gravity separation as in Table 32.

** Cleaning of part of production at Level A cleaning as in Table 31.

the estimated removal would be 8.12×10^6 and 8.73×10^6 short tons sulphur respectively. Hence the total mechanically cleaned coal used for combustion in 1985 would be increased from 190.4×10^6 short tons in the Reference Case to 351.6×10^6 short tons, and the average sulphur content for all combustion coals would be in the range 1.30 to 1.45 per cent. Similarly for the Accelerated Policy Case, the average for all combustion coals would range from 1.30 to 1.44 per cent sulphur.

An alternative strategy would be to clean the coal at two specific gravity separations to redistribute sulphur into a clean low sulphur coal product and a middlings product with higher sulphur. The advantage in doing this is that, by either further cleaning the middlings or applying post-combustion emission control to their combustion, the SO_2 emissions reduction is maximised without the need to progress to the cleaning of the total coal production to lower overall sulphur levels.

Table 32 summarises the estimated sulphur content of coals in 1985 if all coals were processed through a 1.3 specific gravity separation to produce a clean product, followed by a 1.6 specific gravity separation to produce a middlings product. It shows that the clean product (1.3 s.g. float) from each region would contain much lower sulphur content than the middlings (1.3 s.g. sink and 1.6 s.g. float).

In Table 33, the strategies are combined in the best manner to produce two coal products - one for combustion without flue gas desulphurisation (FGD) and the other requiring FGD. This would result in 667.7×10^6 short tons of combustion coals in the 1985 Reference Case having a 1.07 sulphur content and the remaining 189.5×10^6 short tons having 80 per cent of the sulphur removed by FGD giving them an effective average sulphur content of 0.60 per cent.

However, the constraints on achieving these levels of cleaning by 1985 are many. Not only is a greatly increased cleaning plant capacity required but also the quantity of raw coal mined would need to be greater to compensate for the Btu loss during cleaning.

COAL CLEANING - OECD EUROPE

This examination of potential sulphur reduction by cleaning of European coals in 1985 concentrates on the two major producers, Germany (96 million metric tons of hard coal in 1985) and the United Kingdom (135 million metric tons in 1985). Germany will also produce large quantities of lignite (113 million metric tons) in 1985 but, almost 100 per cent of the sulphur in German lignites is organic and so cleaning has no effect on sulphur content. While other countries will produce significant quantities of hard coal and lignite in 1985, little is known of the ratio of pyritic to organic sulphur in the coals, and, therefore, of the cleaning potential.

UK Coal Cleaning Practices

The coal combusted in 1974 in the United Kingdom, with average sulphur content as shown, was supplied to markets as shown in Table 34.

All the coal supplied to the latter 3 sectors was 100 per cent washed coal. The coal for electricity included 17 million metric tons sold as untreated, unwashed run-of-mine coal, 0.5 million metric tons of washed middlings and the remaining 49.5 million metric tons being blended. The blend of washed and untreated coal contained about one-third untreated coal and two-thirds washed coal. It should be remembered that the UK practice is to wash and blend coals to produce desired ash content and heating value. No attention is presently paid to sulphur content.

Table 34

SULPHUR CONTENT OF COMBUSTION COAL IN THE UNITED KINGDOM, 1974

Sector	Coal Combusted (10^6 metric tons)	Average sulphur content (% wt.)(7)
Electricity	67	1.5
Industrial	12	1.6
Domestic	12	1.3
Commercial	3	1.3
Total	94	1.46

From Table 34 the mixture of cleaned and uncleaned coals combusted in 1974 had an average sulphur content of 1.46 per cent. The UK coals generally have an organic sulphur content of 0.8 per cent to 1.0 per cent, and a pyritic sulphur content of 0.3 per cent to 0.8 per cent, with the average sulphur content of raw coal in the range 1.5 per cent to 1.6 per cent.

The 1985 projections show UK coal production to be 135 million metric tons. If that portion for combustion was to be 100 per cent cleaned (no untreated coal) to the sulphur level possible with present practice, then an average sulphur content of coal for combustion in the range 1.2 per cent to 1.3 per cent would be expected. However, without detailed information on the sulphur distribution by size fraction or gravity fraction for all the coals involved, it is difficult to make an accurate evaluation. A reduction in sulphur content of this magnitude indicates a reduction in emissions from coal combustion of about 0.4×10^6 metric tons SO_2 from a potential generation of about 3.0×10^6 metric tons SO_2.

Thus it appears that, while a 13 per cent reduction in emissions could be achieved by 100 per cent cleaning of all coals in 1985, further reduction in sulphur content is possible if the UK coal cleaning plants also tried to minimise sulphur in the final product. In this case, the optimum strategy may be to produce a clean coal product and a middlings by washing at two specific gravities. For example, washing at specific gravities of 1.3 and 1.8 would produce about two-thirds clean product at about 1 per cent sulphur and about one-third middlings at 2.5 per cent to 3.0 per cent sulphur. These figures are rough estimates and a proper assessment would require more detailed information on sulphur distribution in the coal. To effect the further sulphur reduction, the middlings could be crushed and rewashed but it might be preferable to apply flue gas desulphurisation to their combustion products.

FRG Coal Cleaning Practices

In 1974 some 85 per cent of the hard coal production of 101×10^6 metric tons in the Federal Republic of Germany was cleaned. The uncleaned coal included dust below 0.5 mm in size, raw fine coal in the size range 0.5-10.0 mm, and slurry. The quantity of hard coal combusted in each sector and the estimated sulphur contents are shown in Table 35.

Table 35

ESTIMATED SULPHUR CONTENTS OF HARD COAL FOR COMBUSTION IN THE FRG, 1974

Sector	Coal Combusted (10^6 metric tons)	Average Sulphur Content (% wt.)
Electricity	36.6	1.22
Industry	10.6	1.22
Domestic/Commercial/Other	3.3	0.95
Total	50.5	1.20

By 1985, the Federal Republic of Germany plans to clean all coal to produce a clean low sulphur product.

Table 36 summarises typical yields and sulphur contents of the products of coal preparation in the FRG.

The dust, slurry and coarse and fine middlings are likely to be subjected to further cleaning, for example by froth flotation, or used where there is post-combustion emission control. The portion of the coal available for combustion will have an average sulphur content in the range 1.0 per cent to 1.1 per cent, bringing about a

Table 36

TYPICAL QUANTITIES AND SULPHUR CONTENTS OF COAL
PREPARATION IN THE FRG

Product	Quantity %	Sulphur Content (dry) % wt.
Coarse washed coal	22.6	1.10
Fine washed coal	36.8	1.10
Dust	12.7	1.35
Slurry	18.0	1.60
Coarse middlings	4.7	1.80
Fine middlings	5.2	1.70

reduction in SO_2 emissions of about 0.2×10^6 tons SO_2 from the potential generation in 1985. This would represent the maximum potential reduction in SO_2 emissions by the application of present coal cleaning technology.

RESIDUAL FUEL OIL DESULPHURISATION - JAPAN

Hydrodesulphurisation (HDS) of residual fuel oil is a well-established process in Japan. In 1974 heavy oil HDS by-produced about 790,000 tons of elemental sulphur. Japanese refineries use both methods of HDS: indirect HDS in which the vacuum distillate from distillation of heavy oil is desulphurised and then blended back with the vacuum residuum, and direct HDS in which atmospheric residuum is treated directly. Table 37 shows the Japanese HDS capacity installed up to 1974, that planned for installation up to the end of 1977, and that which can be projected for 1985.

Japanese plans for installation of heavy fuel oil desulphurisation capacity beyond 1977 are not clear. A report (10) by the Japanese Environment Agency shows the growth rate in the total HDS and flue gas desulphurisation (FGD) capacities will fall after 1980, with the capacity increasing from 5.0×10^6 in 1980 to only 6.7×10^6 metric tons SO_2 removed in 1985.

In order to examine the potential for further reduction in SO_2 emissions beyond the projected 1985 HDS capacity it is worth noting the capabilities of the HDS processes. Generally, direct desulphurisation of heavy oil can reduce sulphur content by up to 90 per cent for an atmospheric residuum and less for a vacuum residuum, but it is difficult to desulphurise below 1 per cent sulphur. Nonetheless, since 1 per cent sulphur oil has become unsatisfactory for use in

Table 37

JAPANESE HDS CAPACITY

Year	Capacity (10^3 bbl/day)	
	Direct	Indirect*
1973 (actual) (19)	194	668
1978 (planned) (19)	424	957
1985 (projected)	890	1,190

* Vacuum-distillate throughput.

many parts of Japan, several oil companies are building new plants to desulphurise to 0.3 per cent sulphur or less by using new catalysts and two reactors in series. Indirect desulphurisation can remove about 95 per cent of sulphur in the vacuum distillate, down to 0.2 per cent sulphur. Desulphurisation beyond these levels is technically possible but costs begin to rise steeply. Hence, for the processes to be cost effective they are best applied to the highest sulphur residual fuel oils.

Thus, to properly assess the potential for further reduction of residual fuel oil sulphur content it is necessary to know what proportions of the fuel oil pool are of high, medium and low sulphur content. These proportions have been forecast for use in the following example based on the "worst" high sulphur crude oil supply scenario and the expected refinery pattern. Since residual oils high in heavy metal content (particularly vanadium and nickel) are more difficult and more expensive to desulphurise, the following analysis of the desulphurisation potential in Japan has assumed that 10 per cent of high sulphur fuel oil cannot be desulphurised because of such high metal content. Table 38 shows, for the "worst" crude supply 1985 Reference Case, the additional quantity of heavy fuel oil that could be desulphurised and the additional sulphur that would be removed at an 80 per cent removal rate for direct desulphurisation, i.e. desulphurisation of a 2.5 per cent S atmospheric residuum to 0.5 per cent S.

Similar analyses for residual fuel oil in both the "worst" and "best" crude supply scenarios of the Accelerated Policy Case and the "best" supply scenario of the Reference Case show that the projected 1985 HDS capacity will almost be sufficient to desulphurise as much residual fuel oil as available in these cases.

In conclusion, the only significant potential for reducing sulphur below the levels planned for 1985 will be to further desulphurise the product by using additional reactors in series with existing plant or by modifying the existing plant to desulphurise

Table 38

POTENTIAL FOR FURTHER SULPHUR REDUCTION OF RESIDUAL
FUEL OIL IN JAPAN, 1985 REFERENCE CASE - "WORST"
CRUDE SUPPLY

Residual Sulphur Level		Residual Quantity (10^6 metric t.)	Projected HDS Capacity (10^6 metric t.)		Additional Quantity Desulphurised (10^6 metric t.)
			Direct	Indirect*	
High	(3.5% S)	93	34	59	–
Medium	(2.5% S)	50	–	25	25
Low	(0.5% S)	16	–	–	–
Total		159	34	84	25
Total Sulphur Removed (10^6 metric tons)			2.2		0.5

* Expressed as atmospheric residuum throughput.

to lower levels. If environmental effects necessitate this further reduction of sulphur emissions, then these latter options would need to be evaluated, with other technologies for removing sulphur from fuels and combustion gases, as to economic feasibility.

RESIDUAL FUEL OIL DESULPHURISATION - OECD EUROPE

There is little installed hydrodesulphurisation capacity in OECD Europe (6.6×10^6 metric tons per year of indirect capacity and 0.4×10^6 metric tons of direct capacity at the end of 1975), although the utilisation of this capacity is unknown. There are plans to install a further 4.3×10^6 metric tons of indirect capacity in Belgium which, along with other installations forecast, will bring the capacity up to 10.7×10^6 metric tons of indirect (as vacuum distillate throughput) and 4.9×10^6 metric tons of direct HDS by the end of 1980. European plans for installation of HDS capacity beyond 1980 are unknown.

Following the approach used for Japan, estimates have been made of the additional potential for sulphur removal from heavy fuel oil in OECD Europe in 1985 and the direct desulphurisation capacity required to achieve that potential. The results are summarised in Table 39. While it would have been possible to consider the removal by additional indirect HDS capacity, direct HDS has been considered since that process offers the greatest sulphur reduction potential.

Table 39

POTENTIAL FOR FURTHER SULPHUR REDUCTION OF RESIDUAL
FUEL OIL IN OECD EUROPE, 1985

Case	Direct HDS Capacity (10^6 metric tons)	Sulphur Removed (10^6 metric tons)
"Worst" (High S) Crude 1985 Reference Case	198	5.9
"Best" (Low S) Crude 1985 Reference Case	162	4.6
"Worst" (High S) Crude 1985 Accelerated Policy Case	157	4.4
"Best" (Low S) Crude 1985 Accelerated Policy Case	113	3.0

A more detailed analysis of the SO_2 reduction potential by combinations of direct and indirect desulphurisation capacity is presented in Chapter IV. No account has been taken of the small existing capacity.

It should be noted that not all fuel oils will be amenable to desulphurisation for reasons such as high metal content. In addition, residue desulphurisation cannot readily be applied to thermally cracked residue and no vacuum residue is easy to indirectly desulphurise. Allowances have been made for these factors.

As can be seen from the table there is a considerable potential for reducing sulphur levels in residual fuel oils combusted in OECD Europe in 1985 by the installation of additional HDS capacity. In all of the fuel supply cases considered, realisation of this full potential reduction would bring emissions to below their 1974 level. Should the incentive exist to reduce emissions from this source and if HDS is economically competitive with post-combustion control technology (FGD), then considerable reductions in residual fuel oil sulphur content within the region can still be achieved.

RESIDUAL FUEL DESULPHURISATION - NORTH AMERICA

A 1973 estimate (20) put the total United States' hydrodesulphurisation capacity installed or planned to be installed before 1980 at 393,000 bbls/day. Of this, 144,500 bbls/day is for desulphurising heavy gas-oil, 157,000 bbls/day for desulphurisation of vacuum gas-oil and 91,500 bbls/day for direct desulphurisation of residual oils. The sulphur removal by this capacity is 0.25×10^6 long tons of sulphur per year.

As residual fuel oil produced in the United States is largely vacuum residuum, it has a high metal content which makes desulphurisation very difficult. For this reason it has not been possible to estimate the desulphurisation potential of US residual fuel oils as far ahead as 1985. However, much residual fuel oil is imported into the United States, and reduction in the sulphur content of these imports would have benefit, but the potential for further desulphurisation at the Caribbean refineries (from whence a large proportion of imports derive) is limited, as much of their production is already desulphurised. Certainly some desulphurisation potential will exist for other refineries which provide imports and for domestic refineries, and there may be some improvement in overall sulphur levels of fuel oils combusted in the United States by 1985.

In any case, the major contribution to SO_2 emissions in the United States in 1985 will not come from residual fuel oil combustion which will contribute between 4.18×10^6 and 4.73×10^6 metric tons SO_2 to a total between 28.34×10^6 and 30.58×10^6 metric tons SO_2 (potential emissions before application of FGD).

FLUE GAS DESULPHURISATION - OECD EUROPE - POWER PLANTS

During the 19 year period from 1955-1974, the fossil fuel fired electric generation capacity and the consumption of sulphur containing fuels quadrupled as shown in Table 40. Improvements were made in generation efficiency and new technology allowed larger units to be built.

Table 40

HISTORICAL STATISTICS ON POWER PLANTS IN EUROPE

Year	Fossil Fuel Capacity GW	Average Eff. %	Average Load Factor	Average Size New Capacity MW	Sulphur Containing Fuels			Total Fossil Fuels TWh
					Coal TWh	Lignite TWh	Oil TWh	
1955	62	24	0.44	64	179	25	13	235
1960	89	28	0.44	106	229	40	44	340
1965	123	30	0.48	146	303	57	123	517
1970	178	32	0.49	299	341	75	270	685
1974	236	33	0.47	360	313	99	388	974

Oil consumption in power plants increased most significantly from 5.5 per cent in 1955 to 40 per cent in 1974 and as a result accounted for the largest share of electric power production in 1974.

From these statistics, it can be concluded that the electric generation capacity which will be retired between 1974 and 1985 will be predominantly coal fired capacity with unit size being less than

100 MW. It is also noted that the newest capacity which generally operates at a high load factor is the largest capacity. Table 41 shows the characteristics of the 1974 electric power industry.(22)(23) About 10 GW of the total capacity are estimated to be gas turbines. The assumption is made that these are less than 10 years old and less than 30 MW in size.

Table 41

THE SITUATION IN EUROPE, 1974, (6) (22) (23)

Size	Total Capacity GW	Capacity in GW by Age			
		1-8 yrs.	9-15 yrs.	16-20 yrs.	21-30 yrs.
0-30 MW	49	14	8	10	17
30-200 MW	83	11	35	17	20
200-400 MW	59	44	15	-	-
400-600 MW	33	31	2	-	-
over 600 MW	12	12	-	-	-
Total	236	112	60	27	37

The average load factor for 1974 for fossil fuel fired plants was 4.7 per cent or an annual production of 4100 KWh/KW of capacity. The newest capacity (1-8 years old) generally operates at 70 per cent load factor. The remaining age groups in Table 41 were estimated to operate at average load factors of 49 per cent, 30 per cent and 8 per cent.

Using these load factors and subtracting the gas turbine capacity, Table 42 on fuel consumption can be obtained.

Table 42

FOSSIL FUEL CONSUMED BY SIZE AND AGE IN 1974

Size	Total Fuel Consumed TWh	Fuel Consumed in TWh by Age			
		1-8 yrs.	9-15 yrs.	16-20 yrs.	21-30 yrs.
0-30 MW	95	24	34	25	12
30-200 MW	273	67	150	42	14
200-400 MW	334	269	65	-	-
400-600 MW	198	189	9	-	-
over 600 MW	73	73	-	-	-
Total	973	622	258	67	26

It can be noted from Table 42 that significant reduction of sulphur emissions could be realised if flue gas desulphurisation technology were applied to only the large sized power plants of age 1-8 years. In these new, high load factor plants of 200 MW or greater, 55 per cent of the fossil fuel is burned.

By analysing the fuel type of the new capacity installed during these eight years and the change in fuel consumption in power plants, the estimate of fuel consumption in Table 43 was made.

Table 43

EVALUATION OF THE RETROFIT POTENTIAL FOR FGD ON
LARGE, RECENTLY CONSTRUCTED POWER PLANTS

Size	Fuel Consumed in TWh by 1-8 year old Plants				
	Total	Coal	Oil	Gas	Lignite
200-400 MW	269	39	119	49	62
400-600 MW	189	108	51	25	5
over 600 MW	73	57	13	3	0
Total	531	204	183	77	67
Emission Factor (wt. % SO_2)		2.5	5.0	0.0	0.97
Emissions (10^6 tons)	4.8	2.1	1.9	0.0	0.8

From an emissions reduction point of view, the retrofit of SO_2 flue gas control technology appears attractive. However, it is necessary to recognise that retrofit may be difficult and therefore the economics of such an approach must be carefully analysed.

In conclusion, a few additional statistics (23) may be worth noting. In Europe, there are 18 power generation units over 600 MW. These account for 0.7×10^6 metric tons of SO_2 emissions. The 59 power generation units between 400 and 600 MW account for the next 1.7×10^6 metric tons of SO_2 and the 150 power units in the 200-400 MW range account for the remaining 2.4×10^6 metric tons. Since it is estimated that there are over 2,000 power generation units operating in Europe, this is a small fraction.

Table 44 shows the eleven year period from 1974 to 1985, during which the European electric generation capacity is expected to grow modestly. The contrast between the 1985 Reference Case and the Accelerated Policy Case is clearly seen. The production of electricity decreases as conservation measures are implemented and there is a shift to coal and nuclear power from oil.

It is expected that the size of the new generating units installed between 1974 and 1985 will be generally large, continuing the trend of the past. Even though there will be a large nuclear capacity in 1985 which will provide much of the base load, the new fossil fuel capacity will also be needed to operate at high load factors.

Table 44

THE EUROPEAN POWER PLANT FORECAST FOR 1985 (2)

Year	Fossil Fuel Capacity GW	Average Load Factor	Sulphur Containing Fuels			Total Fossil Fuels TWh
			Coal TWh	Lignite TWh	Oil TWh	
1974	236	0.47	313	99	388	974
1980	285	0.45	334	125	511	1,132
1985 Ref	330	0.49	342	137	776	1,417
1985 AP	300	0.46	379	137	524	1,204

Considering these factors, it is estimated that fossil fuel consumption (excluding that consumed by gas turbines) in the 1985 cases will be distributed as shown in Tables 45 and 46. In both 1985 cases, the load factor for the new capacity installed after 1974 is 70 per cent or 6,100 hours. It is technologically feasible to install flue gas control technology on all of this new capacity installed after 1974. If the control technology is planned, designed and constructed with the construction of the power plant, costs will be minimised.

It has been estimated that, because of the installation of replacement and expansion capacity to meet the increased demand for electric power, the emissions from coal, lignite and residual oil-fired generation capacity will increase to $13.3*-15.2 \times 10^6$ metric tons in the 1985 Reference Case and to $10.8*-12.2 \times 10^6$ metric tons in the 1985 Accelerated Policy Case.

The maximum technologically feasible reduction in these emissions would result if all capacity installed after 1974 and all capacity over the size of 200 MW installed after 1967 had flue gas desulphurisation technology. This would require retrofit of capacity installed between 1967 and 1974 during the late 1970s. The result of such an action would mean that there would be installed 98-126 GW of new FGD capacity and 74 GW of retrofit capacity depending on the 1985 forecast.

The emission reduction potential by FGD technology on oil-fired power plants could be 84 per cent (67 per cent from new capacity and 17 per cent from retrofit) and 75 per cent on coal-fired (36 per cent and 39 per cent respectively) for the 1985 Reference Case. For the

---

*) The range depends on the crude supply pattern, particularly the destination of North Sea oil.

1985 Accelerated Policy Case, the potential is 81 per cent from oil (56 per cent and 25 per cent respectively) and 76 per cent from coal (39 per cent and 37 per cent respectively).

Table 45

1985 REFERENCE CASE FOR EUROPE

Size	Total Fuel Consumed TWh	Fuel Consumed in TWh by Age			
		1-5 yrs.	6-11 yrs.	12-19 yrs.	20-30 yrs.
0-30 MW	53	-	-	19	34
30-200 MW	246	49	49	52	96
200-400 MW	513	116	158	209	30
400-600 MW	406	146	109	147	4
over 600 MW	197	79	61	57	-
Total TWh	1,415	390	377	484	164
Average size of unit MW		450	400	350	140

Table 46

1985 ACCELERATED POLICY CASE FOR EUROPE

Size	Total Fuel Consumed TWh	Fuel Consumed in TWh by Age			
		1-5 yrs.	6-11 yrs.	12-19 yrs.	20-30 yrs.
0-30 MW	48	-	-	18	30
30-200 MW	209	24	49	49	86
200-400 MW	450	67	158	197	27
400-600 MW	338	86	110	139	4
over 600 MW	157	43	61	54	-
Total TWh	1,202	220	378	457	147
Average size of unit MW		450	400	350	140

It is recognised, however, that economic, institutional and/or political constraints will most likely preclude any retrofitting of the capacity installed during 1967-1974. However, considering the objectives of this report, it is worthwhile to identify the emission reduction potential.

A Group of Experts on flue gas desulphurisation was asked to forecast the amount of FGD capacity in Europe in 1985. Their estimate was as follows:

	1976 (GW)	1985 (GW)
Austria	0	Possibly some
Federal Republic of Germany	0.18	3.3
Netherlands	0	Possibly some
Scandinavia	0.05-0.15	0.3-0.5
Spain	0	Possibly some
United Kingdom	0.24	Unknown
Remainder of OECD Europe	0	0

The SO_2 emissions reduction resulting from about 4 GW of possible European FGD capacity - 3.5 GW being coal-fired with 1.3 per cent average sulphur and 0.5 GW being oil-fired with 2.5 per cent average sulphur - is estimated to be 210 thousand metric tons SO_2 from coal and 28 thousand metric tons SO_2 from oil. This total represents 1-2 per cent of total emissions from electric power production.

FLUE GAS DESULPHURISATION - OECD NORTH AMERICA - POWER PLANTS

The North American situation shown in Table 47 is dominated by the United States which had about 95 per cent of the 1974 fossil fuel fired generating capacity. The small quantity of fossil fuel fired capacity in Canada is explained by the fact that in 1974, 80 per cent of the total electricity was produced by mostly hydro-power and a little nuclear.

Table 47

HISTORICAL STATISTICS ON POWER PLANTS IN NORTH AMERICA, (6) (22) (23)

Year	Fossil Fuel Capacity GW	Average Eff. %	Average Load Factor	Average Size New Capacity MW	Sulphur Containing Fuels			Total Fossil Fuels TWh
					Coal TWh	Lignite TWh	Oil TWh	
1962	176	35	0.55	300	549	3	58	838
1965	217	35	0.56	450	698	3	81	1,053
1970	311	35	0.56	750	877	9	212	1,513
1974	375	35	0.51	Unknown	993	19	326	1,721

The differences between the electricity generation in North America and in OECD Europe are striking. The average generation efficiency was maintained at 35 per cent in North America for the 12-year period, whereas in Europe it was lower, going from about 29 per cent in 1962 to 33 per cent in 1974. A lower efficiency means more fuel to produce the same quantity of electricity and subsequently more sulphur emissions. In addition, the North American utilities maintained a higher average load factor (0.55 average over the 12 years) than the European utilities (0.48 average). However, within North America, the Canadian average load factor was much lower, about 0.44. The load factor is important in determining the fraction of generating capacity on which it is economically feasible to install flue gas cleaning processes. The annual operating costs are lower for both boiler and flue gas cleaning when load factor is high, since the investment cost and other fixed costs are distributed over more operating hours. The cost of electricity is lower with higher average load factor.

From these statistics, it can be concluded that coal is still the dominant North American fuel for electricity generation. In addition, the electric generation capacity which will be retired between 1974 and 1985 will predominantly be coal-fired with the remainder being gas-fired. The average size of coal-fired boilers installed between 1945 and 1955 is reported to be 100 MW.(24) Therefore, the retired capacity will be at this average size.

Table 48

1974 CANADIAN CAPACITY IN GW BY AGE, (6) (22) (23)

Generating Unit Size	Total Capacity	1-8 yrs.	9-15 yrs.	16-20 yrs.	21-30 yrs.
0-30 MW	2.8	-	1.2	0.5	1.1
30-200 MW	5.5	2.8	1.7	0.7	0.3
200-400 MW	5.1	2.7	2.1	0.2	-
over 400 MW	4.5	4.5	-	-	-
Total	17.8	10.0	5.0	1.4	1.4

If a similar analysis was done on the Canadian utilities as was performed on the European utilities, the same observation would be made from the data in Table 48. That is, it may prove beneficial to SO_2 emissions to install FGD technology to the power plants of age 1-8 years and of 200 MW or greater in size. This is based on the assumption that these plants have the highest load factor and consume the greatest amount of sulphur containing fuels.

Therefore, the potential capacity for flue gas desulphurisation retrofit would be 2.7 GW of electric generation capacity between 200-400 MW in size and 4.5 GW over 400 MW. The available statistics (23) show that for the 4.5 GW, there are eight coal-fired units and one oil-fired unit, and for the 2.7 GW, there are two coal-fired units, one dual coal/gas-fired unit and six units for which the fuel type is unknown. There are several unknown factors that complicate this analysis. These are:

  i) The average Canadian load factor over the 1968-1973 period was 0.44 but fell to 0.37 in 1974. This implies that the new plants (1-8 years in age) probably operated with load factors of around 0.50 only;
  ii) The sulphur contents of combustion coals in Canada vary considerably from low sulphur in Western provinces to high sulphur in the Eastern production and in the imports from the United States;
  iii) Very little is known about the variability of sulphur in Canadian fuel oils.

Table 49

1971 UNITED STATES CAPACITY (24)

Plant Size	No. of Plants	Average Plant Size (MW)	Average Age (years)	Average No. of Boilers per Plant	Sulphur Emitted per Year (as % of Total Emissions)		
					Coal	Oil	Both
0-100 MW	351	39.5	25.9	3.4	3.50	0.35	3.85
100-200 MW	142	145.0	21.2	4.1	4.20	0.86	5.07
200-400 MW	150	290.0	19.3	4.8	13.81	1.63	15.44
400-600 MW	92	494.3	14.4	4.8	15.75	1.82	17.57
600-800 MW	47	683.6	11.4	3.7	10.05	1.18	11.24
800-1000 MW	35	898.1	10.2	4.1	6.73	0.71	7.43
1000-1200 MW	21	1,106.4	11.9	4.5	10.49	0.40	10.89
1200-1400 MW	16	1,272.3	9.5	4.9	8.11	0.37	8.48
1400-1600 MW	7	1,534.0	9.3	5.6	4.88	0.56	5.44
1600-3000 MW	19	1,867.8	6.1	4.5	14.37	0.27	1.63
Total	880				91.89	8.17	100.00

To properly analyse the potential for flue gas desulphurisation retrofit in Canada would require more detailed statistics on a provincial basis.

From the statistics shown in Table 49 the following observations can be made:

1. The coal and oil-fired plants over 750 megawatts (approximately 100 plants) account for 50 per cent of the sulphur dioxide emissions in United States utility industry.

2. The next 100 plants (400-750 megawatts) emit an additional 20 per cent of the sulphur dioxide emissions.
3. The next 15 per cent comes from approximately 180 plants in the range of 150-400 megawatts.

These observations are shown graphically in Figure 1. The report concluded that it is fairly clear that there should be a size determined cut-off point beyond which further sulphur dioxide reductions become increasingly more difficult to achieve. The older the plant, the smaller it is and the smaller the individual boilers. It is to be expected that the cost per kilowatt of installing flue gas desulphurisation on a new boiler increases as the size of the boiler decreases. However, if the boiler is in existence already, it also becomes increasingly more difficult to retrofit a stack gas unit with increase in age and decrease in size.

In 1971, the United States promulgated new source performance standards (NSPS) which apply to all power plants constructed or modified after 17th August, 1971. These standards limit emissions from oil-fired and coal-fired plants respectively to 0.8 lbs. SO_2 per million BTU of heat input and 1.2 lbs. SO_2 per million BTU. Plants constructed or modified before this date are subject to emission regulations adopted by the States as part of regional plans to meet National Ambient Air Quality Standards (NAAQS). Thus the emission regulations for older boilers vary from State to State, and in most cases are not more stringent than the NSPS.

As a result of these standards the flue gas desulphurisation capacity in the United States reached 2.8 GW by the end of 1974. However, many of these early units experienced initial start-up and operational problems, which hampered their use. Most of these were installed on power plant boilers with high load factors (approximately 70 per cent), but because of the operation problems were utilised only roughly 50 per cent of the time that the boiler was operating. During the operation, approximately 75 per cent removal efficiency of SO_2 was the average. Given that these operating parameters are representative, it is estimated that flue gas desulphurisation reduced emissions from coal and oil-fired power plants by 156 thousand metric tons in 1974.

For the eleven year period from 1974 to 1985, the North American fossil fuel fired electric generation capacity is expected to grow as shown in Table 50. During this period, it is estimated that there will be 48 GW of expansion capacity and about 20 GW of new capacity to replace retired capacity.

A group of experts on flue gas desulphurisation was asked to forecast the amount of FGD capacity in North America in 1985. Their estimate was as follows:

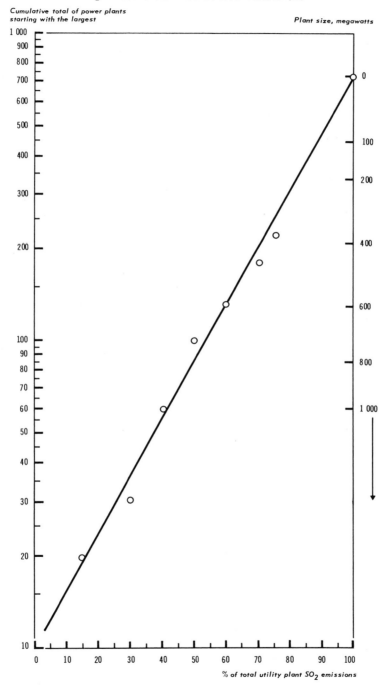

Figure 1
SO₂ EMISSIONS FROM US UTILITY PLANTS (24)

	1976 (GW)	1985 (GW)
United States	6.5	46.4-80.0
Canada	-	3.0- 5.0
OECD - North America	6.5	49.4-85.0

The United States estimate depends on the number of utilities that must retrofit with flue gas desulphurisation and the ability to meet the NSPS with other technologies by 1985.

Table 50

THE NORTH AMERICAN POWER PLANT FORECAST FOR 1985 (2)

Year	Fossil Fuel Capacity GW	Average Load Factor	Sulphur Containing Fuels			Total Fossil Fuels TWh
			Coal TWh	Lignite TWh	Oil TWh	
1974	375	0.51	993	19	326	1,721
1980	404	0.51	1,081	29	348	1,805
1985 Ref.	425	0.56	1,415	42	334	2,086
1985 AP	423	0.54	1,362	41	216	2,017

In the United States, it is expected that the flue gas desulphurisation capacity will be on coal-fired units only burning coal with a 2.0 per cent sulphur average, with a boiler load factor of 0.65 average and an FGD removal efficiency of 90 per cent and reliability of 100 per cent. Then the emissions reduction can be calculated to be 4.3-7.4 million metric tons SO_2. It is also expected that the design, operation and maintenance of FGD units by 1985 will have reached a state of expertise such that 100 per cent reliability during the period of boiler operation will be realisable. An average sulphur content of 2 per cent for coal was used on the basis that the US average coal sulphur content will be about 1.6 per cent S in 1985 but the low sulphur fraction will be used without FGD, leaving a higher sulphur fraction for use with FGD.

In Canada, it is expected that half of the flue gas desulphurisation capacity will be coal-fired units only in the eastern provinces from Ontario eastward, burning high sulphur coals of 3.0 per cent S, with a boiler load factor of 0.55 and an FGD removal efficiency of 90 per cent and reliability of 100 per cent. Then the emissions reduction from coal can be calculated to be 0.18-0.29 million metric tons SO_2. The other half would be on oil-fired boilers with 2.8 per cent S oil. This would produce emissions reduction from oil of 0.09-0.15 million metric tons of SO_2. In comparing these with the two

1985 Canadian emissions forecasts in Tables 11 and 12, it can be concluded that this magnitude of capacity in 1985 is the needed amount to totally control the emissions from electric power production.

If the forecasts of FGD capacity in 1985 are met, the North American electric power producers will have installed the maximum technologically feasible flue gas desulphurisation capacity.

FLUE GAS DESULPHURISATION - JAPAN - POWER PLANTS

During the 13 years from 1962 to 1974, oil clearly became the dominant fuel in electricity generation in Japan, as shown in Table 51. As in most of the OECD world, the consumption of sulphur containing fuels grew severalfold. The Japanese utilities maintained the highest operating load factors in the OECD. During this period, they varied between 0.52 and 0.68.

Table 51

HISTORICAL STATISTICS ON POWER PLANTS IN JAPAN, (6) (22) (23)

Year	Fossil Fuel Capacity GW	Average Load Factor	Average Size New Capacity	Sulphur Containing Fuels			Total Fossil Fuels TWh
				Coal TWh	Lignite TWh	Oil TWh	
1962	14	0.62	150	44	-	31	78
1965	24	0.56	200	51	-	62	113
1970	47	0.67	285	51	-	210	275
1974	78	0.52	500	20	-	301	355

Since 1967, the consumption of coal in electricity generation has been declining to one-fourth of the maximum consumption in 1974. As a result, no coal-fired boilers have been installed since 1970. Most of the old capacity to be retired between 1974 and 1985 will be coal-fired.

The average size of electric generating unit has also increased with 500 MW being the average size installed in 1974. Table 52 characterises the 1974 electric power industry in Japan.

In Japan, flue gas desulphurisation has been used mainly by industries other than the electric power industry. Table 53 is an estimation by MITI of the amount of fuel oil that must be subject to flue gas desulphurisation to attain the Japanese ambient standard of 0.04 ppm SO_2.

In 1974, about 2.5 GW of FGD capacity was installed on power plants and another 3.1 GW on the industrial combustion sources. This capacity would reduce SO_2 emissions by 0.17 million metric tons from power plants and by 0.22 million metric tons from industrial combustion.

Table 52

JAPANESE CAPACITY IN GW BY AGE IN 1974, (6) (22) (23)

Size	Total Capacity	1-8 yrs.	9-15 yrs.	16-20 yrs.	21-30 yrs.
0-30 MW	10	3	2	1	4
30-200 MW	18	6	8	3	1
200-400 MW	31	23	8	-	-
400-600 MW	8	8	-	-	-
600-800 MW	9	9	-	-	-
over 800 MW	2	2	-	-	-
Total	78	51	18	4	5

Table 53

FGD CAPACITY REQUIRED TO ATTAIN 0.04 ppm SO_2

	MW				
	1973	1974	1975	1976	1977
Power industry	776	2,475	3,437	5,649	6,413
Chemical industry	489	1,012	1,920	2,695	3,124
Paper industry	787	1,315	2,040	2,271	2,382
Petroleum industry	236	236	732	1,139	1,303
Metal industry	165	198	253	280	292
Textile industry	61	94	253	506	957
Ceramic industry	110	313	369	446	495
	2,624	5,643	9,004	12,986	14,966

As of the end of 1974, the FGD capacity in operation, under construction and on order amounted to about 20 GW, of which half is from the power industry and the remainder for industrial boilers, sulphuric acid plants, Claus plants and iron ore sintering plants and smelters. The wet lime-limestone scrubbing process of Mitsubishi-JECCO is the most widely used in Japan; more than one-fourth of the 20 GW will use this process.

A significant feature of the Japanese efforts on flue gas desulphurisation is that the processes are designed to produce saleable by-products. This is because Japan has a limited domestic supply of sulphur and sulphur compounds and also limited land which could be used for disposal of useless waste materials. It is reported that about 65 per cent of the FGD capacity produces saleable gypsum, 15 per cent sodium sulphate, 15 per cent sulphuric acid and a remainder of miscellaneous salt products. But there is concern that as both fuel oil and flue gas desulphurisation capacity increase, there will be an oversupply of these products.

For the eleven year period from 1974 to 1985, the Japanese fossil fuel fired electric generation capacity is expected to grow as shown in Table 54.

Table 54

THE JAPANESE POWER PLANT FORECAST FOR 1985 (2)

Year	Fossil Fuel Capacity GW	Average Load Factor	Sulphur Containing Fuels		Total Fossil Fuels TWh
			Coal TWh	Oil TWh	
1974	78	0.52	20	301	355
1980	98	0.57	25	356	488
1985 Ref.	112	0.62	51	319	606
1985 AP	101	0.57	51	216	503

The forecast shows that nuclear, gas and imported coal will account for almost all the increase in electric power generation. In the Accelerated Policy Case, oil consumption by utilities will decrease to 70 per cent of the 1974 levels. Because of a large base loaded nuclear capacity in the Accelerated Policy Case, the load factor for fossil fuel fired capacity in 1985 will probably be lower than in the Reference Case.

With a forecast such as this, there will be little need for additional flue gas desulphurisation capacity to be installed after 1980. A group of experts on flue gas desulphurisation, independently of this forecast, projected that only about 6 GW of additional FGD capacity would be installed in Japan after 1977, making the total FGD capacity 26 GW. If half of this capacity were installed on power plants and half on industrial combustion sources, about 0.9 million metric tons of SO_2 emissions would be eliminated from each sector, or 1.8 million metric tons SO_2 in total. As a result of lower fuel sulphur levels, this full potential reduction would not be achieved for the "best" (low-sulphur) crude supply Accelerated Policy Case.

FLUE GAS DESULPHURISATION - OECD - INDUSTRIAL COMBUSTION

Industrial combustion accounts for a significant fraction of emissions as shown in Table 55. It should be recognised that the fuel oil desulphurisation capacity in North America and Japan will reduce emissions below these levels, but the effect on each sector of use is not known.

Table 55

EMISSIONS FROM INDUSTRIAL COMBUSTION

Million metric tons SO_2

	Hard Coal	Brown Coal	Residual Fuel Oil
Europe - 1974	0.8	0.1	5.5
Europe - 1985 Ref.	0.9	0.2	4.6 - 6.0
N.A. - 1974	2.8	-	1.0
N.A. - 1985 Ref.	4.1	-	2.5
Japan - 1974	-	-	1.3
Japan - 1985 Ref.	-	-	4.8 - 5.7

It is unfortunate that the statistics are not available to permit a detailed analysis of the potential for use of flue gas desulphurisation in this sector. A report from the US EPA (24) analysed the industrial boiler population in 1971 in the United States. From this study they identified 500 coal and oil fired boilers over 300×10^6 BTU/h capacity (30 MW equivalent) which accounted for 50 per cent of the SO_2 emissions. The next 900 boilers, between $100-300 \times 10^6$ BTU/h (10-30 MW equivalent) accounted for 25 per cent of emissions. The remaining 3560 boilers were considered too small for flue gas desulphurisation to be economic.

As mentioned earlier, it is expected that about half of the Japanese flue gas desulphurisation capacity will be on industrial boilers, including many units between 20 and 100 MW equivalent heat rate. Similarly, there will probably be some FGD on industrial boilers in the United States depending upon state or local emissions regulations. No United States Federal regulations for SO_2 emissions apply to industrial boilers as yet.

One interesting observation is that from a chemical process experience point of view, the industrial sector is more suited to specify, design, operate and troubleshoot flue gas desulphurisation processes than the utility sector. To most industries this would be just another chemical process. This raises the question - would it be more practical to require industry to install FGD first and establish the reliability of technology and then require utilities to install it, or is it more beneficial to require the largest emitters to install FGD first?

Chapter IV

ECONOMICS OF DESULPHURISATION

Introduction

The technology for expanding the clean fuel supply and for reducing sulphur emissions beyond the levels possible using the available naturally low-sulphur fuels has been thoroughly examined in the previous chapter. It is without a doubt feasible to use these technologies rather extensively in OECD, as a number of countries are already doing. It is now necessary to examine the relative costs of these technologies in the 1985 scenarios.

The objectives of this chapter are to determine the true sulphur premium for desulphurised oil and coal, and to determine the most economically attractive technologies or combinations of technologies for sulphur removal.

The task of cost analysis is the most difficult of those to be accomplished in this report. More detail of the future is required in cost analysis than in the analysis of the potential for using the different technologies. It is fortunate that this degree of detail could be obtained for OECD Europe and unfortunate that it is not available for the other regions. Therefore this chapter focuses primarily on OECD Europe but certainly the methodology will be applicable and perhaps the results will be qualitatively useful to the other regions of OECD.

All the costs used are for application of technologies in 1985 and have been projected on to a common 1980 cost basis using actual cost escalation figures up to 1977 and assuming an annual cost escalation from 1977 to 1980 of 9 per cent per annum. The costs are estimates and have associated error bands generally ranging from 10 per cent to 30 per cent. This must be recognised when using the absolute cost values. However, the intention of this chapter is to examine the relationship between the costs of different technologies applied in different situations with the size factor, the fuel type and sulphur content, the operating hours and other factors playing a role in determining these relative costs. It is in this cost relationship analysis that this chapter will have its value.

RESIDUAL OIL COMBUSTION

Residual Oil Desulphurisation

The basic cost data for residual oil desulphurisation have been adapted from CONCAWE Report No. 13/72, A Study of the Costs of Residue and Gas Oil Desulphurisation for the Commission of the European Communities.(25) A more recent supplement to the CONCAWE report (26) allowed the costs to be projected on to the 1980 Cost basis, assuming an average crude oil price of ⌀100/metric ton in 1980.

The specific residue desulphurisation cases considered by CONCAWE (25)(26) and used as the basis of this analysis are:

a) direct desulphurisation of high sulphur (4.0 per cent) residue from Kuwait crude oil to 0.5 per cent sulphur,

b) direct desulphurisation of high sulphur (4.0 per cent) residue from Kuwait crude oil to 1.0 per cent sulphur,

c) indirect desulphurisation of high sulphur (4.0 per cent) residue from Kuwait crude oil with a 40 per cent sulphur reduction,

d) direct desulphurisation of medium sulphur (2.5 per cent) residue from Iranian light crude oil to 0.5 per cent sulphur,

e) indirect desulphurisation of medium sulphur (2.5 per cent) residue from Iranian light crude oil with 40 per cent sulphur reduction.

Figures 2, 3 and 4 present the analysis for the "worst" high sulphur crude oil supply scenario of the Reference Case, while Figures 5, 6 and 7 present the same information for the "best" low sulphur crude oil supply scenario of the Reference Case. The letters a) to e) used above have been used to identify the particular cases in the figures.

For each of the cases a) to e), there is a limit to the quantity of sulphur which can be removed, determined in part by the 1985 crude supply scenario and in part by the technology. For example, in case e) the limit is 0.6×10^6 metric tons sulphur in the "worst" crude supply case and similarly 4.6×10^6 metric tons for case a). For the combination of cases (a + e), the limit is the sum or 5.2×10^6 metric tons of sulphur.

For the analysis, the proportions of the fuel oil pool which are of high, medium and low sulphur content have been forecast from the particular crude oil supply scenarios and the expected refinery pattern. Allowance has been made for residues which cannot readily be desulphurised, such as those from thermal cracking, those with a high heavy metals content and vacuum residue.

Figure 2
COST PER TON OF SULPHUR REMOVED VS SULPHUR REMOVAL, «WORST» CASE

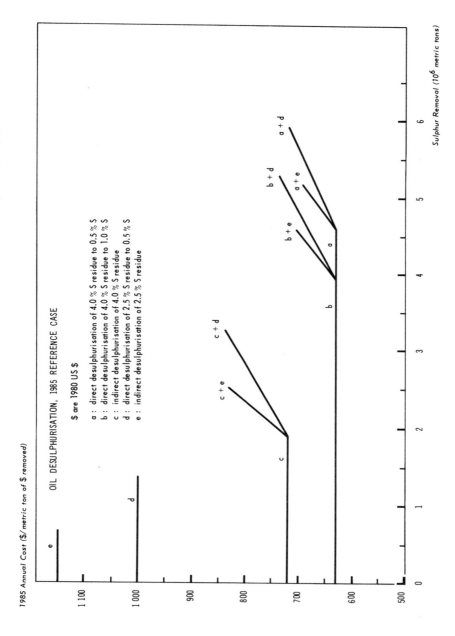

Figure 3
COST PER TON OF OIL TREATED VS SULPHUR REMOVAL, «WORST» CASE

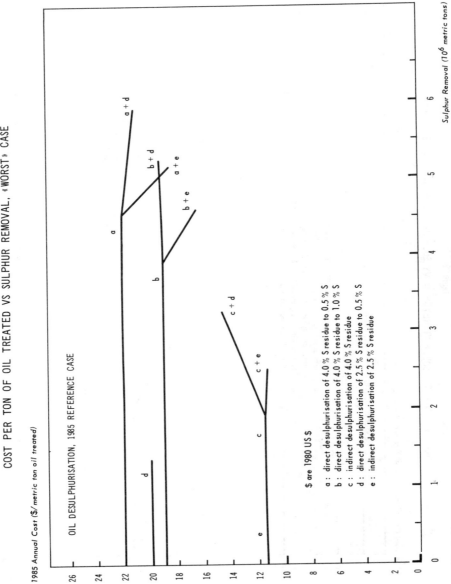

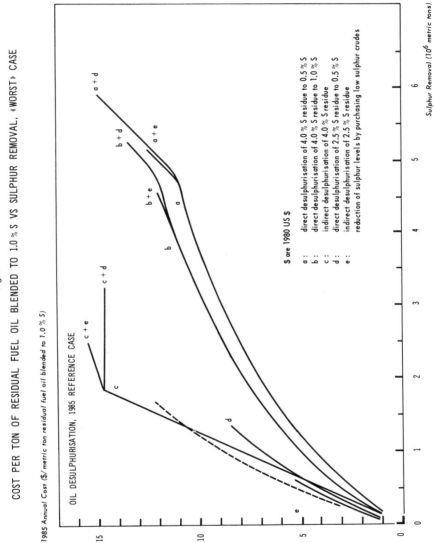

Figure 4
COST PER TON OF RESIDUAL FUEL OIL BLENDED TO 1.0 % S VS SULPHUR REMOVAL, «WORST» CASE

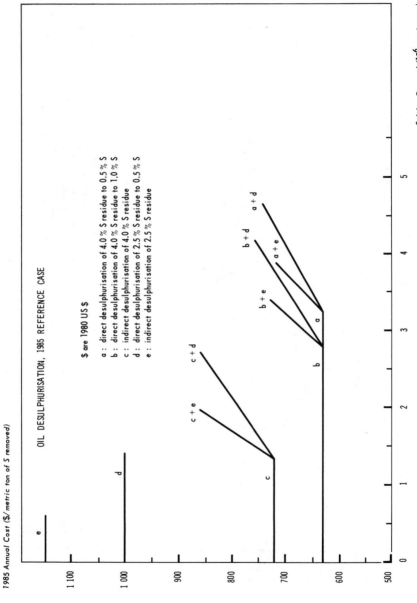

Figure 5
COST PER TON OF SULPHUR REMOVED VS SULPHUR REMOVAL, «BEST» CASE

Figure 6
COST PER TON OF OIL TREATED VS SULPHUR REMOVAL, «BEST» CASE

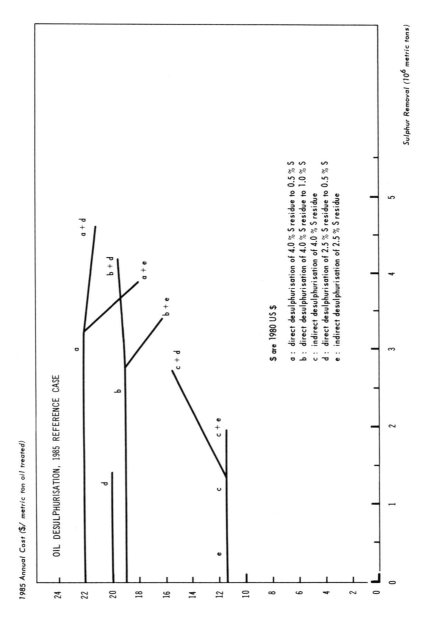

Figure 7
COST PER TON OF RESIDUAL FUEL OIL BLENDED TO 1.0 % S VS SULPHUR REMOVAL, « BEST » CASE

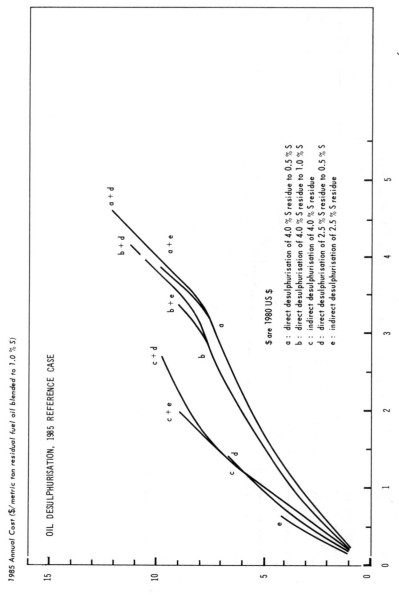

Within the cost ranges given by CONCAWE certain economies of scale have been identified. For example, the average cost differential for a 3,800 metric tons per calendar day direct desulphurisation unit compared with 7,600 metric tons per calendar day unit is +17 per cent. These larger size units can be accommodated by refineries with an annual crude processing capacity of 5×10^6 metric tons or more. In situations where large quantities of sulphur may be removed from the residual oil pool desulphurisation capacity may be installed in refineries smaller than this. However, some 90 per cent, or 903×10^6 metric tons per annum of European refinery capacity installed at the end of 1975 was made up of refineries with a crude processing capacity in excess of 5×10^6 tons/annum. Optimisation of desulphurisation strategies on a regional basis would therefore mean that full advantage could be taken of the economies of scale which have been described here.

The costs for desulphurisation are for specific atmospheric residues but are generally representative for the whole atmospheric residue pool in Europe. Hence the costs should be viewed in terms of the total atmospheric residue pool and not in terms of specific residues, for which costs may be higher or lower than those presented. Costs for desulphurisation of vacuum residues are expected to be 15 to 25 per cent higher than for atmospheric residues.

There are three ways to analyse the cost of residual oil desulphurisation. The first is to examine the actual cost of sulphur removal. Figure 2 shows this. For the residue desulphurisation cases a) to e) and for combinations of these cases, the variation of total annual operating cost per ton of sulphur removed with the quantity of sulphur removed is presented. Mean costs have been used for this graph and there may be a variation of up to $\pm$ 15 per cent on those costs, depending on specific circumstances. The most efficient strategy on a cost per ton of sulphur removed basis can be readily identified. For example, to reduce sulphur levels by 5×10^6 metric tons per annum, it would be more efficient to use a combination of cases a) and d) at ⌀660 per ton of sulphur than to use a combination of cases b) and d) at ⌀720 per ton of sulphur. As the total annual operating costs are the product of the cost per ton of sulphur removed and the quantity of sulphur removed, then the most efficient strategy for sulphur removal is also the one with minimum total cost.

The second way to analyse the costs is to examine how the cost per ton of residual oil treated varies with the quantity of sulphur removed, as shown for each case in Figure 3. The minimum cost strategy identified in Figure 2 has also been shown. Contrasting the information contained in Figures 2 and 3 for the 5×10^6 metric tons per annum sulphur removal already considered, direct desulphurisation of high sulphur residues to 0.5 per cent S and direct desulphurisation of medium sulphur residues has the highest cost increment to the quantity of oil treated (a + d: ⌀22 per ton of oil treated). The lowest cost increment to the quantity of oil treated would result from the indirect desulphurisation of all medium sulphur residues and then direct desulphurisation of high sulphur residues to 0.5 per cent S (e + a: ⌀18 per ton of oil treated). This minimum treatment cost strategy is not that with the lowest total annual operating cost. However, since residual oils are generally blended to meet user needs and specification, it is general practice to assign the costs of treatment of one portion of the blending oils over the cost of the whole blended pool. Therefore, this type of cost per ton of oil treated is not representative of the true cost for blended oil and cannot be compared with the total operating costs in Figure 2. It is therefore worthwhile to look at a third way to analyse the costs.

To include the desulphurisation costs in the cost of the low sulphur fuel oils is the third way of analysing the costs. By blending the naturally low sulphur residue with the desulphurised product and some of the remaining medium or high sulphur oil to produce as much low sulphur residue as possible, at nominally 1.0 per cent sulphur content, and then by distributing desulphurisation costs across this whole low sulphur oil pool, a measure of cost effectiveness in terms of both quantity of sulphur removed and quantity of oil desulphurised is obtained. The total annual operating costs distributed in this way are shown in Figure 4.

Inherent in this redistribution of desulphurisation costs to all the available low sulphur fuel oil is the assumption that market forces will reinforce the premium for low sulphur products. This would not be unreasonable to expect if legislation or other motivation was to enforce sulphur limitations.

It can be seen that the minimum cost strategies which can be identified from Figure 4 are the same as the minimum cost strategies in terms of cost per ton of sulphur removed (Figure 2): i.e. direct desulphurisation of all high sulphur residue to 0.5 per cent S and direct desulphurisation of medium sulphur residue (a + d). To use the same example, to reduce sulphur levels by 5×10^6 metric tons per annum would in reality increase the cost of low sulphur (1% S) fuel oil by ⌀11.90 per metric ton over the cost of medium and high sulphur fuel oil. The true annual cost of desulphurisation in 1985 for Europe in this example could be expressed as ⌀660 per metric ton sulphur removed or as ⌀11.90 incremental cost per metric ton of 1% S fuel oil or as ⌀3.3 billion total cost.

In addition, much desulphurisation (up to 3.8 million metric tons per annum of sulphur removed) can be achieved with the minimum cost strategy at less than the existing premium for low sulphur fuel oils of about $10 per metric ton. For example, the July 1977 price in Western Europe for a 1 per cent S No.6 fuel oil was $84.25-85.50 per metric ton compared with a 3.5 per cent S No.6 fuel oil at $74.50-75.50 per metric ton. (Oil Buyer's Guide.)

Finally, if for their lower sulphur content light, low sulphur crudes are purchased instead of high sulphur crudes, the incremental cost should be assigned to the resulting low sulphur residue pool. Then the cost difference between the "worst" high sulphur crude oil supply scenario and the "best" crude oil supply scenario can be calculated and has been added to Figure 4. The price differential was $13.80 per metric ton, based on the October 1977 prices of heavy Middle Eastern crudes compared with light, North African, West African or North Sea crudes. The conclusion to be drawn from this cost comparison in Figure 4 is that, in most cases, residual fuel desulphurisation is a cheaper way of reducing sulphur levels in the fuel oil supply than is the purchase of low sulphur crude oils. However, as the price differential of the lighter, low sulphur crudes could be distributed over a wider product range by the oil industry, this could make purchase of low sulphur crudes cost less than residual oil desulphurisation in some cases.

In conclusion, consider that residual oil desulphurisation is the only technology used to augment the clean fuel supply and to reduce sulphur emissions. The optimal cost strategy would be first to desulphurise all high sulphur residue to 0.5 per cent S and next to do the same to the medium sulphur residue, blending the products with the naturally occurring low sulphur residue and distributing the cost over the resulting clean fuel oil pool. This may often cost less than the purchase of additional low sulphur crude oil.

Flue Gas Desulphurisation

In the United States and in Japan, lime and limestone scrubbing is clearly the most prevalent and preferred flue gas desulphurisation process. It is expected that in Europe the lime and limestone processes will also be the most prevalent, since the large amount of experience with these systems will have sizeable benefits in terms of costs and reliability. For the purpose of estimating the cost of flue gas desulphurisation in Europe, it has been assumed that limestone scrubbing, at 90 per cent removal with sludge fixation and disposal is representative of the typical scrubber in Europe.

To estimate the 1985 annual operating costs for FGD in 1980 currency, the oil-fired power plants which are suitable for FGD were divided into three age categories and four size categories. These are:

Age categories
- I. New power plants in the planning stage
- II. New power plants in the design, construction or start up stage (easy retrofit)
- III. Existing power plants installed after 1966 (moderate to difficult retrofit)

Size categories

- 30-200 MW
- 200-400 MW
- 400-600 MW
- over 600 MW

For the two crude supply cases of the 1985 Reference Case, the 1985 annual operating costs for FGD in Europe are shown in Figure 8. These costs assume that no sulphur is removed by residual oil desulphurisation. The lowest cost is for new power plants of category I, over 600 MW. The cost of each step from lowest to highest corresponds to the following combination of size and age:

	Age Category	Size Category
Lowest	I	over 600
	I	400-600
	II	over 600
	I	200-400
	II	400-600
	II	200-400
	I	30-200
	III	over 600
	III	400-600
	II	30-200
Highest	III	200-400

The equations used to estimate the FGD costs are given in Table 56. The basic cost data from which these were derived have been taken from a US EPA/TVA report (27) on cost estimates for five FGD processes. The investment costs in this report are in 1974 dollars and have been escalated by a factor of 1.68 to arrive at 1980 dollars. The operating costs were updated in a subsequent paper (28) by the TVA and the 1974-1977 trends in raw material, labour and utilities costs are the basis for estimating the 1980 operating costs. For the European utility, a higher rate of capital charges has been used than was used in the TVA report. The sludge fixation and disposal cost was estimated to be ∅16/short ton of dry waste in 1980. For these costs, it is assumed that the FGD systems are not first-of-a-kind systems or those requiring large amounts of R & D or trouble-shooting.

As an example of the use of Figure 8, consider the strategy to remove 2.0 million metric tons of sulphur by FGD. This would involve installing FGD on all new power plants in category I over 200 MW and

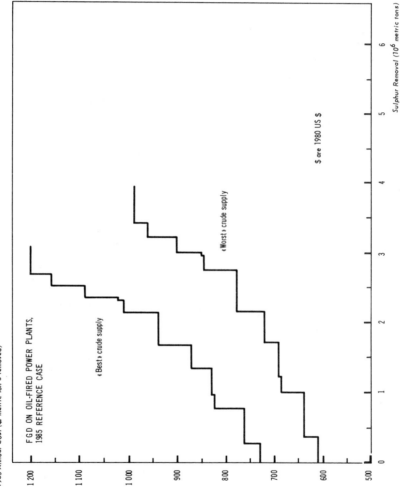

Figure 8
COST PER TON OF SULPHUR REMOVED VS SULPHUR REMOVAL, «BEST» AND «WORST» CASES

Table 56

SUMMARY OF FGD COST EQUATIONS

Basis: Limestone Scrubbing, 1980 Cost Basis (1.68 escalation over 1974 on capital investment; raw materials, utilities and fuel oil escalation vary based on 1974-1977 trends), 90 per cent S removal, with sludge fixation and disposal, with flue gas reheat, and, for coal, with particulate control to 0.1 lb/million Btu of heat input.

FGD on New Coal-Fired Power Plant (particulate control)

Average Annual Operating Cost; Size in MW, H in hours/yr operation.

$$\cancel{\$} = \left(\frac{\text{Size}}{500}\right)^{0.7} \left[6708 + 676(\%S)^{0.7}\right] + \left(\frac{\text{Size}}{500}\right)\left[0.664 + 0.1939\,(\%S)\right](H)$$

Sulphur removed (metric tons/yr) = $1.457\,(\%S)(H)\left(\frac{\text{Size}}{500}\right)$

FGD on Existing Coal-Fired Power Plant (particulate control)

$$\cancel{\$} = \left(\frac{\text{Size}}{500}\right)^{0.7}(R)\left[6708 + 676(\%S)^{0.7}\right] + \left(\frac{\text{Size}}{500}\right)\left[0.859 + 0.1984\,(S\%)\right](H)$$

R is 1.2 for easy retrofit and 1.4 for difficult retrofit

Sulphur removed (metric tons/yr) = $1.490(\%S)\,(H)\left(\frac{\text{Size}}{500}\right)$

FGD on New Oil-Fired Power Plant (no particulate control)

$$\cancel{\$} = \left(\frac{\text{Size}}{500}\right)^{0.7}\left[3924 + 563\,(\%S)^{0.7}\right] + \left(\frac{\text{Size}}{500}\right)\left[0.484 + 0.1512\,(\%S)\right](H)$$

Sulphur removed (metric tons/yr) = $0.995\,(\%S)\,(H)\left(\frac{\text{Size}}{500}\right)$

FGD on Existing Oil-Fired Power Plant (no particulate control)

$$\cancel{\$} = \left(\frac{\text{Size}}{500}\right)^{0.7}(R)\left[3924 + 563(\%S)^{0.7}\right] + \left(\frac{\text{Size}}{500}\right)\left[0.569 + 0.1552\,(\%S)\right](H)$$

Sulphur removed (metric tons/yr) = $1.017\,(\%S)\,(H)\left(\frac{\text{Size}}{500}\right)$

Figure 9
COST PER TON OF SULPHUR REMOVED VS SULPHUR REMOVAL, «WORST» CASE

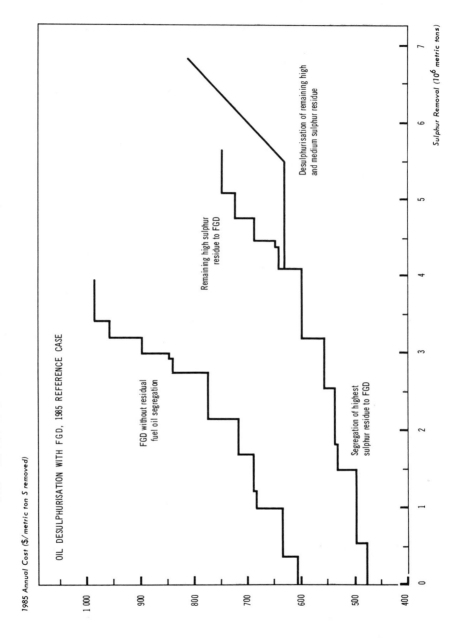

all new power plants in category II over 400 MW. The operating cost for all these FGD units would range from 610-720 $/metric ton of sulphur removed ("worst" crude supply case).

To contrast the costs for the "best" and "worst" crude supply, consider the example of using FGD to reduce 1985 emissions to 20×10^6 metric tons of SO_2. To accomplish this would require FGD to remove approximately 1×10^6 metric tons of sulphur in the "best" case and 2.7×10^6 metric tons in the "worst" case. The number of power plants required to install FGD in each case is clearly different but the ranges of costs incurred are very similar.

It can be seen that the costs of FGD for sulphur removal of up to 2.0 million metric tons are comparable to the costs of residual oil desulphurisation cases a) and b) in Figure 2 (direct desulphurisation of 4.0 per cent S residue). However, above this amount of sulphur removal, residual oil desulphurisation appears to have cost advantages.

Segregation of High-Sulphur Fuel Oil to FGD

Lower costs for emission control in terms of $/metric ton of sulphur removal can be expected if the technologies of FGD and residual oil desulphurisation are each applied to a separate fuel oil pool. This implies segregation of the high and medium sulphur residual fuel oils into a pool of highest sulphur content for use in oil-fired power plants with FGD and into a pool of the remaining residue for direct desulphurisation to 0.5 per cent S.

For the European oil-fired power plants, the high sulphur pool would have an average sulphur content of 4.0 per cent and consist of about 100×10^6 metric tons in the 1985 "worst" case. This would be sufficient residual fuel oil to supply all power plants of age categories I and II over 200 MW in size. These would then be equipped with limestone flue gas desulphurisation processes as described earlier. Approximately 4×10^6 metric tons of sulphur would be removed annually at operating costs in 1985 ranging from $475 to $600 per metric ton of sulphur removed (1980 dollars). This is shown in Figure 9 by the lower curve.

For comparison, the cost from Figure 8 of FGD without segregation of high sulphur fuel oil is included in Figure 9. As expected, segregation of the high sulphur fuel oil to power plants with FGD reduces the cost per ton of sulphur removed. The total annual operating cost is about the same in both cases but with segregation much more sulphur can be removed.

At the point of 4×10^6 million metric tons of sulphur removal, the whole highest sulphur pool, consisting of 100×10^6 metric tons of 4.0 per cent sulphur fuel oil, is consumed and only the pool consisting of about 110×10^6 metric tons of high sulphur residue

(35 per cent) and medium sulphur residue (65 per cent) remains. As Figure 9 shows, the costs are lower if this is treated by direct residual oil desulphurisation than if it is used with FGD units on the smaller (30-200 MW) and/or older oil-fired power plants.

Therefore, it would be most cost effective in terms of cost per ton of sulphur removed for the European electric power industry to install FGD units on all oil-fired power plants installed after 1974 and over 200 MW in size and for the oil industry to segregate the high and medium sulphur fuel oils into two pools, with direct desulphurisation of the lower sulphur pool.

COAL COMBUSTION

Coal Cleaning

As the physical properties of coal and coal cleaning practices and costs vary considerably from one coal producing country to another within the OECD European region, an economic analysis of coal cleaning in the region as a whole would not be sufficiently representative for the coals of any one country. In addition, coal properties and, therefore, cleaning potential are best documented for the two major producers, the Federal Republic of Germany and the United Kingdom. By 1985 Germany plans to clean all coal to produce a clean product having an average sulphur content of about 1.0 per cent. The present coal cleaning practice in the United Kingdom is not oriented specifically towards sulphur reduction but towards producing a coal of uniform heating value and low ash content. Taking the United Kingdom as an example for Europe, costs have been analysed for desulphurisation by coal cleaning.

In considering the current UK coal cleaning practice there is potential for reorienting this practice to increase sulphur removal. Three levels of cleaning can be examined:

a) cleaning to produce a single product coal of uniform heating value and ash content,
b) cleaning at two specific gravity separations to redistribute sulphur into a clean low-pyritic sulphur coal product and a middlings product with higher sulphur, and
c) redistribution as in b), cleaning of the middlings to reduce sulphur and blending with clean coal product.

The redistribution of sulphur into the two product streams does not in itself reduce the overall sulphur content of the coal. It is through further processing of the middlings, for example by regrinding, more washing and froth flotation, that the SO_2 emissions reduction can be achieved. Alternatively, the middlings may be used where FGD is applied to the post-combustion gases. The costs of these two alternatives may be compared.

The costs for a new coal cleaning plant for each of these levels of cleaning have been estimated by a group of experts, as shown in Table 57. The operating costs in this table include only operation and maintenance. In the analysis, it is necessary to add the annualised capital charges at the rate of 20 per cent of the capital investment per annual ton of capacity, which for 3000 hours per year of operation are ∅1.70 per metric ton. The 1976 operating cost would be ∅2.80 per metric ton raw coal (∅4.70 per metric ton washed coal).

In 1974, the UK coal combustion was 94×10^6 metric tons of which about 60×10^6 metric tons were cleaned. In 1985 the coal combustion is forecast to be slightly higher. The annual operating cost in 1985 to produce coal of uniform heating value and ash content is ∅6.10 per metric ton of washed coal (escalated to a 1980 dollars basis). To redistribute the sulphur in the coal (level b) or reduce sulphur further (level c) would increment the annual operating cost of ∅6.10 per metric ton by the costs shown in Table 58. The total cost of ∅7.50-8.25 per metric ton would represent the sulphur premium for desulphurised coal (level c) over unwashed coal and the cost of ∅1.40-2.15 would represent the premium over washed coal (level a).

It has been estimated that washing UK coal at specific gravities of 1.3 and 1.8 would produce about one-third middlings at 2.5 per cent to 3.0 per cent sulphur and about two-thirds clean product at about 1.0 per cent sulphur. As the organic sulphur content of the coal (0.8 to 1.0 per cent) would be almost the same for the clean product and the middlings, the pyritic sulphur content of the middlings would be in the range of 1.5 per cent to 2.0 per cent sulphur. It is expected that further processing could reduce the middlings sulphur content to 1.8 per cent with a minimum loss of coal to the waste. The ultimate coal sulphur content, if clean coal and middlings were blended after all processing, would be about 0.25 per cent sulphur lower.

From Table 58 the incremental cost of sulphur removal (level c) over washed coal (level a) is in the range ∅560-860 per metric ton of sulphur. However, if existing cleaning practice requires that only some percentage of coal be cleaned, then to the incremental cost would have to be added a share of the base cost for coal washing. For example, at present 50 per cent of UK coals for electricity sector combustion are washed. To increase sulphur removal on these 50 per cent would cost ∅560-860 per metric ton sulphur removed. If the other 50 per cent of the coal were to be treated for sulphur removal a 50 per cent share of the base cost would be included, making the cost for all 100 per cent ∅1,800-2,100 per metric ton of sulphur removed.

Table 57

COSTS FOR NEW COAL CLEANING PLANTS

Currency at December 1976 values*

	Capital Costs (currency/metric ton of raw coal/hr)		Operating Costs (currency/metric ton of raw coal)	
	FRG	UK	FRG	UK
Base cost to produce coal of uniform heating value and ash content	30-40,000 DM ($12-16,000)	£15,000 ($25,000)	2.5-3 DM ($1.00-1.20)	70 p ($1.17)
Incremental cost over base to redistribute pyritic sulphur into product and middlings	+ 12% (assumes dense medium cyclones at SG = 1.30-1.32)		5-16%	
Incremental cost over base for redistribution and reduction of sulphur by secondary processing of middlings	25-32%		20-40%	

* $1.00 = £0.60 = 2.50 DM

Table 58

SULPHUR REMOVAL AND INCREMENTAL OPERATING COSTS
OVER BASE COST FOR REDISTRIBUTING PYRITIC
SULPHUR AND REPROCESSING MIDDLINGS, UNITED KINGDOM

	Incremental Operating Cost ($/metric ton washed coal)*	Sulphur Removed (% wt. S)	Incremental Cost of Sulphur Removal ($/metric ton)*
Redistribute sulphur into product and middlings	0.55-0.85	0	
Redistribution and secondary processing of middlings	1.40-2.15	0.25	560-860

* $ are 1980 US $

In conclusion, for the countries which wash a large percentage of combustion coal production, a cleaner low sulphur product can be obtained at an incremental cost of $1.40-2.15 per metric ton of coal or at a sulphur removal cost of $560-860 per metric ton of sulphur. Sulphur removal costs increase rapidly if a lower percentage is generally washed. For Europe in 1985, if all combustion coal could be reduced by 0.25 per cent sulphur content on average by this method, the quantity of sulphur removed would be 0.5-0.6×10^6 metric tons. The potential removal is limited, but coal cleaning may be the most economic method for coal desulphurisation.

Flue Gas Desulphurisation

The basis used for flue gas desulphurisation costs for oil-fired power plants has also been used for coal-fired power plants, with the addition of particulate control to 0.1 lb/million Btu.* The basis and the cost equations are presented earlier in Table 56. The operating costs per unit by size, age and solid fuel type have been determined in Table 59. The costs shown assume sulphur contents for each solid fuel as follows:

- hard coal: the European average of 1.35 per cent S in the 1985 Reference Case,
- middlings: 2.75 per cent S, within the range expected for washing UK coals at specific gravities of 1.3 and 1.8 to produce clean product and middlings,
- lignite: the European average of 1.15 per cent S in the 1985 Reference Case.

In determining the sulphur reduction, the sulphur normally retained in the ash has not been included as part of the reduction. This is consistent with the estimation of the potential reductions in Chapter III.

As expected, the costs are much higher for FGD on hard coal-fired power plants because of the relatively low sulphur content of European hard coals. The FGD costs for new plants ranging from ∅1,039-1,783 per metric ton of sulphur removed are much higher than coal cleaning costs, ∅560-860 per metric ton of sulphur removed. However, in the case of UK coals for electricity sector combustion where only 50 per cent are washed, the cost of coal cleaning, ∅1,800-2,100 per metric ton of sulphur removed, is higher. It is also clear that retrofit of FGD on existing coal-fired plants would rarely be economical.

For lignite combustion, the costs of FGD are much more attractive. Since lignites are often very low in pyritic sulphur content and cannot be desulphurised by washing, FGD is the only option for sulphur removal. Since over 80 per cent of European lignites will be consumed in power plants in 1985, the potential exists to remove 1.2×10^6 metric tons SO_2 from new lignite-fired power plants and 0.7×10^6 metric tons SO_2 from retrofit of existing lignite power plants.

---

* Metric equivalent is 0.18 kg/kcal.

Table 59

FGD OPERATING COST PER UNIT FOR SOLID FUEL FIRED POWER PLANTS, 1985 REFERENCE CASE

¢/metric ton S removed*

Unit Size	Power Plants Installed 1981-1985			Power Plants Installed 1975-1980			Power Plants Installed 1967-1974		
	Hard Coal	Middlings	Lignite	Hard Coal	Middlings	Lignite	Hard Coal	Middlings	Lignite
150	1,511	848	738	1,783	988	856	1,848	1,022	886
300	1,203	685	600	1,420	796	694	1,666	926	804
500	1,099	630	553	1,298	732	639	1,560	870	756
700	1,039	598	526	1,228	695	607			

* ¢ are 1980 US ¢.

Coal Cleaning vs Flue Gas Desulphurisation

For hard coal, it may be more economical to use a combination of coal cleaning and FGD where the coal cleaning is used to segregate a middlings product for use in power plants with FGD.

The incremental cost of segregation of washed coal into clean coal and middlings is ∅0.55-0.85 per metric ton of coal washed from Table 58. Similarly the incremental cost of segregation of an unwashed coal is ∅6.65-6.95 per metric ton of coal washed. Depending upon the 1985 practice in a country for using washed or unwashed coal, the proper proportion of these costs would represent the segregation cost.

After segregation, two options exist for the middlings. The first is to subject the middlings to further processing at the cost of ∅0.85-1.30 per metric ton of washed coal. The second is to use the middlings in a power plant with FGD at operating costs ranging from ∅598-988 per metric ton of sulphur removed from new power plants, or ∅14-24 per metric ton of coal combusted.

Table 60 summarises the total cost for these two options for two cases of different washing practices. For the option of segregation with further processing of middlings the costs per ton of coal in the table are simply the total cost of segregation and further processing of the middlings distributed over only the cleaned coal product (approximately two-thirds of the products).

For the other option of segregating middlings for use with FGD, it is necessary to have a price differential between the cleaned coal and the middlings which would be sufficient to offset the cost of FGD. In this case, the cleaned coal product must also bear the cost of segregation. The cost per ton of coal in the table represents the price margin above the base cost for washed coal necessary to establish the price differential. For the example in Table 60, this price differential would be set at ∅14-24 per ton of coal.

It can be seen from Table 60 that, if the existing practice is to wash a high proportion of power plant coal, then the incremental cost per ton of sulphur removed for improved sulphur removal in the coal cleaning plant is much less than the cost of producing high sulphur middlings for use with FGD. If the existing practice is to wash a lower proportion of power plant coal then the reverse may be the case. However, it should be remembered that, in all cases, FGD has the potential to remove much larger amounts of sulphur than does the coal cleaning.

If, in Europe in 1985, all power plants greater than 100 MW capacity constructed since 1975 were to use segregated middlings, then the sulphur removal would be 1.4×10^6 metric tons from the combustion of about 60×10^6 metric tons of middlings. The total annual operating cost for such a strategy is in the range ∅1,100-

Table 60

INCREMENTAL COST FOR SULPHUR REMOVED BY COMBINED COAL WASHING AND FGD

Normal Practice for Combustion Coal	Coal Segregation and Processing of Middlings (no FGD)		Coal Segregation with Middlings to FGD	
	(¢/metric t. clean coal product)*	(¢/metric t. S)*	(¢/metric t. coal)*	(¢/metric t. S)*
100 % washed coal	1.25-1.95	560-860	6.00-10.20	770-1,300
50 % washed, 50 % unwashed coal	5.80-6.55	1,800-2,100	9.10-13.20	1,160-1,680

* 1985 annual operating costs in 1980 US ¢

$2,400 \times 10^6$. In comparison, the sulphur removal from segregation and further processing of middlings for all combustion coals would be only 0.5–0.6×10^6 metric tons.

Another cost analysis of coal cleaning with scrubbing for sulphur control was carried out by the US EPA.(29) The report examined a number of case studies combining some physical coal cleaning with cleaning of some of the combustion gases to meet the US Federal standard of 1.2 lbs/MBTU for SO_2 emissions. It was concluded that, in many cases, the net cost of physical coal cleaning followed by scrubbing of part of the flue gas to meet standards is substantially less than that associated with using only a full scale scrubbing system. However, this conclusion depends upon cost benefits from using clean coal, such as increased heat content, transportation savings, ash disposal savings and pulverising savings. Whether this approach would have the same cost benefit in Europe would therefore be dependent on whether the benefits of using cleaned coal are already being realised. In countries where a significant proportion of coal is not already cleaned the combined approach may have a cost benefit.

In conclusion, it is apparent that for most European hard coals, coal cleaning to reduce sulphur content will be the most economical approach either because of lowest incremental cost or because of other benefits derived from the use of clean coal. However, if further sulphur reduction is needed beyond that which is possible with coal cleaning, a high sulphur middlings should be segregated for use in a power plant with FGD.

SUMMARY

If the maximum reduction of total SO_2 emissions is desired the following European approach would have the lowest cost of sulphur removal (1980 dollars):

- segregate high sulphur and medium sulphur residual fuel oils,
- install flue gas desulphurisation in all oil-fired power plants over 200 MW and constructed since 1974 and use the high sulphur fuel oils ($475-630/metric tons of S removed),
- desulphurise by direct residue desulphurisation the remaining high and medium sulphur residual fuel oils to a level of 0.5 per cent S ($630-810/metric ton of S removed),
- physically wash all hard coals to minimise sulphur content ($560-2,100/metric ton of S removed),

- install FGD on all lignite-fired boilers over 100 MW and constructed since 1967 (∅520-890 metric ton of S removed),
- require that all imported coals be washed to minimise sulphur content,
- use naturally low sulphur or cleaned fuel in the domestic, commercial and small industrial sector where FGD is not practical.

This approach would remove approximately 13×10^6 tons of SO_2 from residual fuel oil combustion in 1985 at a total operating cost in 1985 of approximately ∅4 billion. It would also remove approximately 1×10^6 tons of SO_2 from hard coal combustion at an operating cost of approximately ∅0.35 billion and approximately 2×10^6 tons SO_2 from lignite combustion at ∅0.6 billion. In the 1985 "worst" case, SO_2 emissions would be reduced from 25.4×10^6 metric tons of SO_2 to approximately $9-10 \times 10^6$ metric tons at an annual operating cost of ∅5 billion.

If it is desired to maintain emissions at their present level of 20 million metric tons of SO_2 rather than to obtain the maximum emission reduction, a standstill approach could be put into practice by 1985. In the "worst" 1985 case this would mean a reduction of 6×10^6 metric tons SO_2. The washing of all hard coals could reduce SO_2 emissions by 1×10^6 metric tons at an annual 1985 operating cost of approximately ∅0.35 billion. Installation of FGD on all new (post 1974) lignite-fired boilers over 100 MW would reduce SO_2 emissions by another 1×10^6 metric tons at ∅0.3 billion in 1985. The remaining 4×10^6 metric tons reduction could be accomplished by segregation of 4.0 per cent S fuel oil to new power plants (post 1980) with FGD or by direct desulphurisation of high sulphur residual oil to 0.5 per cent S. (The cost of low sulphur fuel oil would be incremented by ∅7/metric ton.) The 1985 operating cost would range from ∅1.0-1.25 billion for these two options. The SO_2 emissions could be reduced by only 3×10^6 metric tons by the purchase of additional low sulphur imported crude oil at an additional cost of about ∅1.75 billion.

A standstill approach which would require the removal of about 6×10^6 metric tons of SO_2 in 1985 would result in a 1985 annual operating cost of ∅1.55-1.80 billion.

BIBLIOGRAPHY

1. The OECD Programme on Long Range Transport of Air Pollutants - Measurements and Findings, Organisation for Economic Co-operation and Development, Paris, 1977.

2. World Energy Outlook, Organisation for Economic Co-operation and Development, Paris, 1977.

3. Energy Statistics 1973-1975, Organisation for Economic Co-operation and Development, Paris, 1976.

4. Statistics of Energy 1960-1974, Organisation for Economic Co-operation and Development, Paris, 1975.

5. 1974 Oil Statistics - Supply and Disposal, Organisation for Economic Co-operation and Development, Paris, 1976.

6. Energy Balances of OECD Countries, 1973-1975, Organisation for Economic Co-operation and Development, Paris, 1977.

7. Report and Conclusions of the Joint Ad Hoc Group on Air Pollution from Fuel Combustion in Stationary Sources, Organisation for Economic Co-operation and Development, Paris, 1973.

8. Compilation of Air Pollution Emission Factors, Second Edition, AP-42, US Environmental Protection Agency, Research Triangle Park, North Carolina, 1976.

9. The Sulphur Grid Method, No. 9/75, CONCAWE Ad Hoc Group on Desulphurisation/Sulphur Legislation, Stichting CONCAWE, The Hague, December, 1975.

10. An Econometric Model for a Long-term Environmental Preservation Plan, Planning and Co-ordination Bureau, Environment Agency, Japan, October, 1976.

11. Federal Power Commission News, Vol. 7 and 8, US Federal Power Commission, Washington, D.C., 1974-1975.

12. US Energy Outlook - A Report of the National Petroleum Council's Committee on US Energy Outlook, December 1972.

13. 1973 National Emissions Report, EPA-450/2-76-007. US Environmental Protection Agency, Research Triangle Park, North Carolina, May 1976.

14. Sulfur Reduction Potential of the Coals of the United States, RI 8118, J.A. Cavallaro, M.T. Johnson, A.W. Deurbrouck, US Bureau of Mines, Pittsburgh, Pa., 1976.

15. 1975 Report on Coke Production, US Bureau of Mines, Washington, D.C., 1976.

16. Effects of Air Quality Requirements on Coal Supply, US Bureau of Mines, Washington, D.C., May 1976.

17. National Energy Outlook, FEA-N-75/713, Federal Energy Administration, Washington, D.C., February 1976.

18. Mineral Industry Surveys, Coal - Bituminous and Lignite in 1974, US Bureau of Mines, Washington, D.C., January 27, 1976.

19. General Survey Paper from the Technical Forum on Flue Gas Desulfurisation, Technical Transfer Institute, Tokyo, 1977.

20. Census of Oil Desulfurisation to Achieve Environmental oals, Paper No. 19-C, R.M. Jimeson and L.W. Richardson, American Institute of Chemical Engineers 4th Joint Meeting with the Canadian Society of Chemical Engineers, Vancouver, British Columbia, September 11, 1973.

21. Recommendation of the Council on Measures for Further Air Pollution Control, OECD, C(74)219, Paris, 25 November 1974.

22. The Electricity Supply Industry, 23rd Enquiry, Organisation for Economic Co-operation and Development, Paris, 1974.

23. The World Market for Electric Power Equipment, University of Sussex, Science Policy Research Unit, 1972.

24. Evaluation of R & D Investment Alternatives for SO_x Air Pollution Control Processes, EPA-650/2-74-098, September 1974.

25. A Study of the Costs of Residue and Gas Oil Desulphurisation for the Commission of the European Communities, Report No. 13/72, Stichting CONCAWE, The Hague, December, 1972.

26. Supplement to Report No. 13/72, A Study of the Costs of Residue and Gas Oil Desulphurisation for the Commission of the European Communities, R.J. Ellis, Stichting CONCAWE, The Hague, November, 1974.

27. Detailed Cost Estimates for Advanced Effluent Desulfurisation Processes, US EPA-600/2-75-006 or TVA Bulletin Y-90, Washington, D.C., January 1975.

28. Flue Gas Desulfurisation Economics, G.G. McGlamery et al., TVA, Muscle Shoals, Alabama, March 1976.

29. Coal Cleaning with Scrubbing for Sulphur Control: An Engineering/Economic Summary, EPA-600/9-77-017, US Environmental Protection Agency, August, 1977.

OECD SALES AGENTS
DÉPOSITAIRES DES PUBLICATIONS DE L'OCDE

ARGENTINA – ARGENTINE
Carlos Hirsch S.R.L., Florida 165,
BUENOS-AIRES, Tel. 33-1787-2391 Y 30-7122

AUSTRALIA – AUSTRALIE
International B.C.N. Library Suppliers Pty Ltd.,
161 Sturt St., South MELBOURNE, Vic. 3205. Tel. 699-6388
P.O.Box 202, COLLAROY, NSW 2097. Tel. 982 4515

AUSTRIA – AUTRICHE
Gerold and Co., Graben 31, WIEN 1. Tel. 52.22.35

BELGIUM – BELGIQUE
Librairie des Sciences,
Coudenberg 76-78, B 1000 BRUXELLES 1. Tel. 512-05-60

BRAZIL – BRÉSIL
Mestre Jou S.A., Rua Guaipá 518,
Caixa Postal 24090, 05089 SAO PAULO 10. Tel. 261-1920
Rua Senador Dantas 19 s/205-6, RIO DE JANEIRO GB.
Tel. 232-07. 32

CANADA
Renouf Publishing Company Limited,
2182 St. Catherine Street West,
MONTREAL, Quebec H3H 1M7 Tel. (514) 937-3519

DENMARK – DANEMARK
Munksgaards Boghandel,
Nørregade 6, 1165 KØBENHAVN K. Tel. (01) 12 69 70

FINLAND – FINLANDE
Akateeminen Kirjakauppa
Keskuskatu 1, 00100 HELSINKI 10. Tel. 625.901

FRANCE
Bureau des Publications de l'OCDE,
2 rue André-Pascal, 75775 PARIS CEDEX 16. Tel. 524.81.67
Principal correspondant :
13602 AIX-EN-PROVENCE : Librairie de l'Université.
Tel. 26.18.08

GERMANY – ALLEMAGNE
Verlag Weltarchiv G.m.b.H.
D 2000 HAMBURG-36, Neuer Jungfernstieg 21.
Tel. 040-35-62-500

GREECE – GRÈCE
Librairie Kauffmann, 28 rue du Stade,
ATHÈNES 132. Tel. 322.21.60

HONG-KONG
Government Information Services,
Sales and Publications Office, Beaconsfield House, 1st floor,
Queen's Road, Central.   Tel. H-233191

ICELAND – ISLANDE
Snaebjörn Jónsson and Co., h.f.,
Hafnarstraeti 4 and 9, P.O.B. 1131, REYKJAVIK.
Tel. 13133/14281/11936

INDIA – INDE
Oxford Book and Stationery Co.:
NEW DELHI, Scindia House. Tel. 45896
CALCUTTA, 17 Park Street. Tel.240832

IRELAND - IRLANDE
Eason and Son, 40 Lower O'Connell Street,
P.O.B. 42, DUBLIN 1. Tel. 74 39 35

ISRAËL
Emanuel Brown: 35 Allenby Road, TEL AVIV. Tel. 51049/54082
also at:
9, Shlomzion Hamalka Street, JERUSALEM. Tel. 234807
48, Nahlath Benjamin Street, TEL AVIV. Tel. 53276

ITALY – ITALIE
Libreria Commissionaria Sansoni:
Via Lamarmora 45, 50121 FIRENZE. Tel. 579751
Via Bartolini 29, 20155 MILANO. Tel. 365083
Sub-depositari:
Editrice e Libreria Herder,
Piazza Montecitorio 120, 00 186 ROMA. Tel. 674628
Libreria Hoepli, Via Hoepli 5, 20121 MILANO. Tel. 865446
Libreria Lattes, Via Garibaldi 3, 10122 TORINO. Tel. 519274
La diffusione delle edizioni OCSE è inoltre assicurata dalle migliori librerie nelle città più importanti.

JAPAN – JAPON
OECD Publications Center,
Akasaka Park Building, 2-3-4 Akasaka, Minato-ku,
TOKYO 107. Tel. 586-2016

KOREA - CORÉE
Pan Korea Book Corporation,
P.O.Box n°101 Kwangwhamun, SÉOUL. Tel. 72-7369

LEBANON – LIBAN
Documenta Scientifica/Redico,
Edison Building, Bliss Street, P.O.Box 5641, BEIRUT.
Tel. 354429-344425

MEXICO & CENTRAL AMERICA
Centro de Publicaciones de Organismos Internacionales S.A.,
Av. Chapultepec 345, Apartado Postal 6-981
MEXICO 6, D.F. Tel. 533-45-09

THE NETHERLANDS – PAYS-BAS
Staatsuitgeverij
Chr. Plantijnstraat
'S-GRAVENHAGE. Tel. 070-814511
Voor bestillingen: Tel. 070-624551

NEW ZEALAND – NOUVELLE-ZÉLANDE
The Publications Manager,
Government Printing Office,
WELLINGTON: Mulgrave Street (Private Bag),
World Trade Centre, Cubacade, Cuba Street,
Rutherford House, Lambton Quay, Tel. 737-320
AUCKLAND: Rutland Street (P.O.Box 5344), Tel. 32.919
CHRISTCHURCH: 130 Oxford Tce (Private Bag), Tel. 50.331
HAMILTON: Barton Street (P.O.Box 857), Tel. 80.103
DUNEDIN: T & G Building, Princes Street (P.O.Box 1104),
Tel. 78.294

NORWAY – NORVÈGE
Johan Grundt Tanums Bokhandel,
Karl Johansgate 41/43, OSLO 1. Tel. 02-332980

PAKISTAN
Mirza Book Agency, 65 Shahrah Quaid-E-Azam, LAHORE 3.
Tel. 66839

PHILIPPINES
R.M. Garcia Publishing House, 903 Quezon Blvd. Ext.,
QUEZON CITY, P.O.Box 1860 – MANILA. Tel. 99.98.47

PORTUGAL
Livraria Portugal, Rua do Carmo 70-74, LISBOA 2. Tel. 360582/3

SPAIN – ESPAGNE
Mundi-Prensa Libros, S.A.
Castelló 37, Apartado 1223, MADRID-1. Tel. 275.46.55
Libreria Bastinos, Pelayo, 52, BARCELONA 1. Tel. 222.06.00

SWEDEN – SUÈDE
AB CE Fritzes Kungl Hovbokhandel,
Box 16 356, S 103 27 STH, Regeringsgatan 12,
DS STOCKHOLM. Tel. 08/23 89 00

SWITZERLAND – SUISSE
Librairie Payot, 6 rue Grenus, 1211 GENÈVE 11. Tel. 022-31.89.50

TAIWAN – FORMOSE
National Book Company,
84-5 Sing Sung Rd., Sec. 3, TAIPEI 107. Tel. 321.0698

UNITED KINGDOM – ROYAUME-UNI
H.M. Stationery Office, P.O.B. 569,
LONDON SEI 9 NH. Tel. 01-928-6977, Ext. 410
or
49 High Holborn, LONDON WC1V 6 HB (personal callers)
Branches at: EDINBURGH, BIRMINGHAM, BRISTOL,
MANCHESTER, CARDIFF, BELFAST.

UNITED STATES OF AMERICA
OECD Publications Center, Suite 1207, 1750 Pennsylvania Ave.,
N.W. WASHINGTON, D.C.20006. Tel.   (202)724-1857

VENEZUELA
Libreria del Este, Avda. F. Miranda 52, Edificio Galipán,
CARACAS 106. Tel. 32 23 01/33 26 04/33 24 73

YUGOSLAVIA – YOUGOSLAVIE
Jugoslovenska Knjiga, Terazije 27, P.O.B. 36, BEOGRAD.
Tel. 621-992

Les commandes provenant de pays où l'OCDE n'a pas encore désigné de dépositaire peuvent être adressées à :
OCDE, Bureau des Publications, 2 rue André-Pascal, 75775 PARIS CEDEX 16.
Orders and inquiries from countries where sales agents have not yet been appointed may be sent to:
OECD, Publications Office, 2 rue André-Pascal, 75775 PARIS CEDEX 16.

OECD PUBLICATIONS, 2, rue André-Pascal, 75775 Paris Cedex 16 - No. 40 711 1978
PRINTED IN FRANCE